Smart Sensor Networks Using AI for Industry 4.0

Advances in Intelligent Decision-Making, Systems Engineering, and Project Management

This new book series will report the latest research and developments in the field of information technology, engineering and manufacturing, construction, consulting, healthcare, military applications, production, networks, traffic management, crisis response, human interfaces, and other related and applied fields. It will cover all project types, such as organizational development, strategy, product development, engineer-to-order manufacturing, infrastructure and systems delivery, and industries and industry sectors where projects take place, such as professional services, and the public sector including international development and cooperation etc. This new series will publish research on all fields of information technology, engineering, and manufacturing including the growth and testing of new computational methods, the management and analysis of different types of data, and the implementation of novel engineering applications in all areas of information technology and engineering. It will also publish on inventive treatment methodologies, diagnosis tools and techniques, and the best practices for managers, practitioners, and consultants in a wide range of organizations and fields including police, defense, procurement, communications, transport, management, electrical, electronic, and aerospace requirements.

Blockchain Technology for Data Privacy Management
Edited by Sudhir Kumar Sharma, Bharat Bhushan, Aditya Khamparia, Parma Nand Astya, and Narayan C. Debnath

Smart Sensor Networks Using AI for Industry 4.0
Applications and New Opportunities
Edited by Soumya Ranjan Nayak, Biswa Mohan Sahoo, Muthukumaran Malarvel, and Jibitesh Mishra

For more information about this series, please visit: https://www.routledge.com/ Advances-in-Intelligent-Decision-Making-Systems-Engineering-and-Project-Management/book-series/CRCAIDMSEPM

Smart Sensor Networks Using AI for Industry 4.0

Applications and New Opportunities

Edited by

Soumya Ranjan Nayak,
Biswa Mohan Sahoo,
Muthukumaran Malarvel, and
Jibitesh Mishra

CRC Press
Taylor & Francis Group
Boca Raton London New York

CRC Press is an imprint of the
Taylor & Francis Group, an **informa** business

First edition published 2022
by CRC Press
6000 Broken Sound Parkway NW, Suite 300, Boca Raton, FL 33487-2742

and by CRC Press
2 Park Square, Milton Park, Abingdon, Oxon, OX14 4RN

CRC Press is an imprint of Taylor & Francis Group, LLC

Library of Congress Cataloging-in-Publication Data
Names: Nayak, Soumya Ranjan, 1984- editor.
Title: Smart sensor networks using AI for industry 4.0 : applications and new opportunities / edited by Soumya Ranjan Nayak, Biswa Mohan Sahoo, Muthukumaran Malarvel, and Jibitesh Mishra.
Description: First edition. | Boca Raton : CRC Press, 2022. | Series: Advances in intelligent decision-making, systems engineering, and project management | Includes bibliographical references and index.
Identifiers: LCCN 2021018338 (print) | LCCN 2021018339 (ebook) | ISBN 9780367702120 (hbk) | ISBN 9780367702137 (pbk) | ISBN 9781003145028 (ebk)
Subjects: LCSH: Industry 4.0. | Artificial intelligence. | Wireless sensor networks.
Classification: LCC T59.6 .S63 2022 (print) | LCC T59.6 (ebook) | DDC 658.4/038028563--dc23
LC record available at https://lccn.loc.gov/2021018338
LC ebook record available at https://lccn.loc.gov/2021018339

ISBN: 978-0-367-70212-0 (hbk)
ISBN: 978-0-367-70213-7 (pbk)
ISBN: 978-1-003-14502-8 (ebk)

DOI: 10.1201/9781003145028

Typeset in Times
by SPi Technologies India Pvt Ltd (Straive)

Contents

Preface

The applications of the wireless sensor network (WSN) increase rapidly due to three basic goals such as availability, confidentially, and integrity. WSN is a collection of a large number of sensor nodes. Each sensor node contains a battery having limited energy capacity. The purpose of this node is to receive, process, and send the data and information. Based on this purpose, there are several challenges in WSN, such as deployment, design constraints, energy constraint, limited bandwidth, node costs, and security. The factors that influence these challenges are coverage, dependability, range, reliability, scalability, security, speed, etc. It causes several types of uncertainties and imprecise information. Hence, there is a need for some innovative, intelligent, nature-inspired techniques in the area of WSN, so that the aforementioned issues are estimated efficiently. Therefore, the proposed book effectively helps academicians, researchers, computer professionals, industry people, and valued users.

The influence and recent impact of the wireless sensor network using artificial intelligence on modern society, technology, and the telecom sector are remarkable. Many activities would not be performed without the sensor network, which has become such an important part of current research and technology. WSN is a truly interdisciplinary subject that draws from synergistic developments involving many disciplines, and it is used in medical diagnosis, telecommunication, computer network vision, and many other fields.

Gaining high-level understanding from WSN is a key requirement for in-depth analysis. One aspect of study that is assisting with this advancement is AI in the sensor network. The new sciences like 5G, energy hole problem, optimized multipath routing protocol, multiple sinks, energy clustering, etc., have gained momentum and popularity, as they have become key topics of research in the area of wireless sensor network. This book has put a thrust on this vital area.

This is a text for use in a first practical course in implementation of wireless sensor network using recent artificial intelligence technology and its analysis, for final-year undergraduate or first-year postgraduate students with a background in biomedical sensor engineering, computer intelligence, remote sensing, radiologic sciences, or physics. Designed for readers who will become 'end users' of WSN in various domains, it emphasizes the conceptual framework and the effective use of WSN tools and uses mathematics as a tool, minimizing the advanced mathematical development of other textbooks.

Featuring research on topics such as maximum lifetime of sensor networks, energy hole problem, no-inspired algorithm, efficient clustering analysis, and AI-based WSN, this book is ideally designed for system engineers, computer engineers, professionals, academicians, researchers, and students seeking coverage on problem-oriented processing techniques and sensor technologies. The book is an essential reference source that discusses wireless sensor network applications and analysis, including optimization technique, AI-based approaches, node clustering, and network life span, as well as recent trends in other evolutionary approaches.

This book is intended to give the recent trends on sensor analysis using AI for network life span, node clustering, and many more related applications, and to understand and study different application areas. It focuses mainly on stepwise discussion, exhaustive literature review, detailed analysis and discussion, rigorous experimentation results, and application-oriented approach.

Matter: this book contains some artificial intelligence algorithms for optimizing several issues of WSN. The main aim of this book is to solve or innovate different problems of WSN in terms of several applications.

About the Editors

Soumya Ranjan Nayak is Assistant Professor at Amity School of Engineering and Technology, Amity University, Noida, India. He received his PhD degree in computer science and engineering under MHRD Govt. of India fellowship from CET, BPUT Rourkela, India; with preceded degree of MTech and BTech degrees in computer science and engineering. He has published over seventy articles in peer-reviewed journals and conferences of international repute like Elsevier, Springer, World Scientific, IOS Press, Taylor & Francis Group, Inderscience, IGI Global, etc. Apart from that, he has 12 book chapters, six books, three Indian patents (one granted), and two international patents to his credit. His current research interests include medical image analysis and classification, machine learning, deep learning, pattern recognition, fractal graphics and computer vision. His publications have more than 400 citations, an h-index of 12, and an i10 index of 15 (Google Scholar). He serves as a reviewer of many peer-reviewed journals such as *IEEE Journal of Biomedical and Health Informatics*, *Applied Mathematics and Computation*, *Journal of Applied Remote Sensing*, *Mathematical Problems in Engineering*, *International Journal of Light and Electron Optics*, *Journal of Intelligent and Fuzzy Systems*, *Future Generation Computer Systems*, and *Pattern Recognition Letters*. He has also served as Technical Program Committee Member of several conferences of international repute.

Biswa Mohan Sahoo is a Senior IEEE member and received his BTech and MTech degrees in computer science and engineering from Biju Patnaik University of Technology, Odisha, India, and his PhD degree in computer science and engineering from the Indian Institute of Technology (ISM), Dhanbad, India. He is currently Assistant Professor at Amity University, Uttar Pradesh, India. He has published more than 15 articles in prestigious international peer-reviewed journals and conferences in the area of wireless sensor networks, swarm intelligence, and image processing. He currently focuses on artificial intelligence approaches on sensor networks. His research interest area is the wireless sensor network with evolutionary algorithm and artificial intelligence.

Muthukumaran Malarvel is a researcher and academician working as Associate Professor in the Centre for Research and Innovation Network (CURIN), Institute of Engineering and Technology, Chitkara University, Punjab, India. His 17+ years of experience includes the information technology industry, teaching, and rich research experience. He received a doctoral degree in computer science and engineering and was awarded a Senior Research Fellowship (SRF) under the funded project of Board of Research in Nuclear Sciences (BRNS), Government of India. He has published more than 20 research articles indexed in SCI, SCIE, and SCOPUS journals and conferences. He is an editor and reviewer of reputed journals and has also edited three books with a reputable publisher. He has filed four intellectual property patents in India and received one patent grant. His research interests are digital image processing, pattern recognition, quantum image processing, and machine learning techniques.

Jibitesh Mishra has interests in research, consultancy, working on new technologies, and projects. He has continuously worked for the College of Engineering and Technology, Biju Patnaik University of Technology, Rourkela, India, since 1994. After completing his PhD in 2001, he has completed projects like Orissa Tourism Information Systems. One of his papers "On calculation of fractal dimension of images" has been cited by many researchers. His first book, *Design of Information Systems: A Modern Approach*, was widely published in China as well. He previously worked in function point metrics in association with Infosys Ltd. He has written another book, *L-systems Fractal*, published by Elsevier. He started introducing the concept of web engineering in India in 2006 after organizing the International Conference on Web Engineering and Applications (ICWA 2006). On the social front, he coordinated the training program for SC & ST graduate engineers in 2008. He was on sabbatical leave during 2008–2010 at King Khalid University, Abha, Saudi Arabia. He wrote a book on e-commerce published by Macmillan during that period which was later purchased by Laxmi Publisher. Currently he is head of the Department of Computer Science and Application in CET. His current research initiatives are in the areas of data analytics, engineering for mobile apps, and ontology engineering.

Contributors

Haritha Venkata Naga Siva Sruthi Addanki
Department of Computer Science and Engineering
Koneru Lakshmaiah Education Foundation
Guntur, Andhra Pradesh, India

Ambuj Kumar Agarwal
Chitkara University Institute of Engineering and Technology
Chitkara University
Punjab, India

Rakesh Ahuja
Chitkara University Institute of Engineering and Technology
Chitkara University
Punjab, India

Ch. Balaswamy
Gudlavalleru Engineering College
Gudlavalleru, Andhra Pradesh, India

Saira Bano
Department of Computer Science and Engineering
Vel Tech University
Chennai, Tamil Nadu, India

K. Baskaran
Alagappa Chettiar Government College of Engineering and Technology
Karaikudi, Tamil Nadu, India

Shelly Bhardwaj
St. Soldier Institute of Engineering and Technology
Jalandhar, Punjab, India

Jayashree Dev
College of Engineering and Technology
Bhubaneswar, Odisha, India

T. Ganesan
Department of Computer Science and Engineering
Koneru Lakshmaiah Education Foundation
Guntur, Andhra Pradesh, India

Amara SA L G Gopala Gupta
Department of Computer Science and Engineering
Koneru Lakshmaiah Education
Guntur, Andhra Pradesh, India

Ravichander Janapati
SR University
Warangal, Telangana, India

Ravi Kumar Jatoth
National Institute of Technology
Warangal, Telangana, India

Ramandeep Kaur
St. Soldier Institute of Engineering and Technology
Jalandhar, Punjab, India

A. M. Senthil Kumar
Department of Computer Science and Engineering
Koneru Lakshmaiah Education Foundation
Guntur, Andhra Pradesh, India

Jibitesh Mishra
College of Engineering and Technology
Bhubaneswar, Odisha, India

Soumya Ranjan Nayak
Amity School of Engineering and Technology
Amity University Uttar Pradesh
Noida, Uttar Pradesh, India

Vipin Pal
National Institute of Technology
Meghalaya, India

G. Palai
Department of Electronics and
    Communication Engineering
GITA
Bhubaneswar, Odisha, India

Nabanita Paul
International Institute of Information
    Technology
Bangalore, Karnataka, India

Yamini Pemmasani
Department of Computer Science and
    Engineering
Koneru Lakshmaiah Education
    Foundation
Guntur, Andhra Pradesh, India

Nihar Ranjan Pradhan
Department of Computer Science and
    Engineering
National Institute of Technology
Meghalaya, India

G. Syam Prasad
Department of Computer Science and
    Engineering
Narasaraopeta Engineering College
Guntur, Andhra Pradesh, India

Purnima
Department of Computer Science
Sri Guru Gobind Singh College
Panjab University
Chandigarh, Punjab, India

Pothuraju Rajarajeswari
Department of Computer Science and
    Engineering
Koneru Lakshmaiah Education
    Foundation
Guntur, Andhra Pradesh, India

S. Ramalingam
Alagappa Chettiar Government College
    of Engineering and Technology
Karaikudi, Tamil Nadu, India

A. Brahmananda Reddy
VNR VJIET College
Hyderabad, Telangana, India

Ranjeet Kumar Rout
National Institute of Technology
Srinagar, Jammu and Kashmir, India

Biswa Mohan Sahoo
Amity University
Uttar Pradesh, India

Ramesh Chandra Sahoo
Utkal University
Bhubaneswar, Odisha, India

Gurpreet Singh Saini
St. Soldier Institute of Engineering and
    Technology
Jalandhar, Punjab, India

Biswaranjan Sarangi
Biju Patnaik University of Technology
Odisha, India

Manish Sharma
Chitkara University Institute of
    Engineering and Technology
Chitkara University
Punjab, India

Pawan Kumar Sharma
National Institute of Technology
Meghalaya, India

Roop Lal Sharma
St. Soldier Institute of Engineering and
    Technology
Jalandhar, Punjab, India

Akhilendra Pratap Singh
Department of Computer Science and
  Engineering
National Institute of Technology
Meghalaya, India

Jaspreet Singh
National Institute of Technology
Meghalaya, India

Vasantha Sravani
Department of Computer Science and
  Engineering
Koneru Lakshmaiah Education
  Foundation
Guntur, Andhra Pradesh, India

Dama Srinu
Department of Computer Science and
  Engineering
Koneru Lakshmaiah Education
  Foundation
Guntur, Andhra Pradesh, India

Umashankar Subramaniam
College of Engineering
Prince Sultan University
Riyadh, Saudi Arabia

Ayusee Swain
Department of Electrical Engineering
College of Engineering and Technology
Bhubaneswar, Odisha, India

K. P. Swain
Department of Electronics and
  Communication Engineering
GITA
Bhubaneswar, Odisha, India

Abhinav Tomar
Netaji Subhas University of Technology
Delhi, India

Biswajit Tripathy
GITA
Bhubaneswar, Odisha, India

Yogita
National Institute of Technology
Meghalaya, India

1 Optimization of Wireless Sensor Networks using Bio-Inspired Algorithm

Ayusee Swain

College of Engineering and Technology, Bhubaneswar, Odisha, India

K. P. Swain and G. Palai

GITA, Bhubaneswar, Odisha, India

Soumya Ranjan Nayak

Amity University, Noida, Uttar Pradesh, India

CONTENTS

DOI: 10.1201/9781003145028-1

1

1.1   INTRODUCTION

A wireless sensor network (WSN) is a self-configured and infrastructure-less wireless network consisting of distributed sensor nodes and sink nodes, which are deployed in large numbers to monitor various physical and environmental conditions in real time, like temperature, pressure, motion, or vibration, and pass their data from sensor nodes to the base station where these data are aggregated, processed, and analyzed. The sensor nodes communicate among themselves and the sink node using radio signals. A WSN is equipped with sensors, transceivers, computing devices, and power components. The base station or sink node acts as an interface between the network and end-users. Apart from a wide range of applications in various fields like IoT, healthcare, military, surveillance, threat detection, agriculture, and the industrial sector, WSNs also have several issues and challenges, like limited power supply, storage, computational capacity, performance, network throughput, energy efficiency, security issues, node deployment, routing, packet delivery, and scalability. Lifetime maximization of networks is a major concern in WSN, especially when a large number of sensor nodes are deployed in a field where maintenance is quite difficult. Therefore, optimization of WSN is essential in order to reduce redundancy and energy consumption and to maximize network lifetime.

Optimization is the process of creating the best solution with respect to a set of prioritized constraints for real-world problems. The main purpose of the optimization algorithm is to maximize or minimize the objective function by comparing various candidate solutions so as to find the optimum results. The Wireless Sensor Network optimization is required for achieving major goals like minimized energy consumption and maximized network lifespan.

Two different types of optimization algorithms are widely used today: (1) deterministic algorithms and (2) stochastic algorithms. A deterministic algorithm always produces the same output when a particular input is passed through it any number of times. On the other hand, stochastic algorithm techniques are equipped with random components that find a different solution in each run even with a similar starting point. In the deterministic approach, the chances of local optima stagnation is high due to its dependency on the initial solution, whereas the stochastic approach avoids local optima and increases the chances of finding the global optimum. This is the main reason why stochastic optimization algorithms have recently been very popular with a wide range of applications.

Stochastic algorithms are classified into two different classes: (1) individual-based and (2) population-based algorithms. Individual-based algorithms start with one solution which is improved and evaluated until the objective function reaches the global maxima or minima. Their algorithms are more prone to local optima stagnation or premature convergence. On the other hand, a population-based algorithm employs a set of initial solutions and improves them constantly to eventually estimate the global optimum. It provides a greater exploration of the search landscape and hence lowers the probability of local optima stagnation. Because of this reason, population-based algorithms are commonly used to solve real-world problems.

Population-based algorithms are divided into three classes: (1) evolutionary algorithms, (2) physics-based algorithms, and (3) swarm-based algorithms. Evolutionary algorithms are inspired by the laws of natural evolution. Examples of these algorithms are genetic algorithms, genetic programming, and biogeography-based optimization. Physics-based algorithms imitate the physical laws of the universe; for example, gravitational search algorithms, central force optimization, and particle collision algorithm. Swarm-based algorithms imitate the social behaviors of a group of animals, for example, particle swarm optimization, grey wolf optimization, and ant colony optimization.

The population-based bio-inspired optimization algorithm is an emerging approach that is based on the principles of the biological and ecological mechanisms in nature. These algorithms imitate biological nature in order to deal with real-world problems. The strategies and computations used in these algorithms are often simple, robust, adaptive, and flexible. There are many bio-inspired computing techniques in the literature for solving a wide range of problems. Evolutionary algorithms and swarm-based algorithms are two important and popular classes in bio-inspired algorithms. In this study, we have analyzed three bio-inspired algorithms—namely, the genetic algorithm (GA), ant colony optimization (ACO) algorithm, and particle swarm optimization (PSO) algorithm—and implemented them in WSN optimization. The GA and ACO algorithms are used to find the shortest path for routing of data packets from source to destination in a WSN in order to balance energy and data load in the network. These two optimizing techniques are then compared based on their ease of implementation and time complexity to solve the optimized routing problems. The PSO algorithm is used to find an optimized position for the cluster head in a cluster of nodes, which can aggregate and process the data transmitted by member nodes and forward it to the base station, thus minimizing the overall energy consumption by the nodes and increasing lifespan of the network.

1.2 LITERATURE REVIEW

The different algorithms are implemented pertaining to localizing issues, clustering, cluster head selection, energy consumption by nodes, optimal path selection, various environment constraints, data gathering, security, network lifetime, data compression, and self-organizing; these are now the current focus area of research in WSNs. Mostly the broad area of research is allocated in three major concerns: energy usage, quality of service, and security management [1]. Among many algorithms, we focus majorly on three algorithms—genetic algorithm, ant colony optimization, and particle swarm optimization—which are implemented widely in the sensor network.

Some review kinds of literature are discussed in [2–7] which discourses the different algorithms that are used in wireless sensor networks (WSNs). The node localization issues are highlighted [2] with the application and challenges related to WSNs. This also emphasizes the different categories of the algorithm with the improved version of the soft computing technique to overcome the concerns related to the sensor network. The various metaheuristic algorithms for the localization issues pertaining to the different nodes of a sensor network are discoursed in [3]. The image compression algorithms which are implemented in a WSN are explained in [4], where different research problems along with their potentials are the main target. The design issues in WSNs along with a state-of-the-art review related to the routing protocol are proposed in [5], and also some future trends regarding the secure routing of the sensor network are highlighted. The exploration, information, and survey about multipath routing protocols that are used in a WSN are discussed in [6] along with the analysis of security issues related to the common attack in the network. Future guidelines are proposed in [7] regarding the multichannel routing protocols for sensor networks. It also discusses the relevant pros and cons of related challenges in routing which may be treated as a road map for future research.

Considering the genetic algorithm (GA) in the field of WSNs, some works of literature are discussed in [8–16]. By balancing the computational load, an optimized GA is proposed in [8] to locate the unknown load which is also improved the lifetime of nodes. In [9], HQCA (high-quality clustering algorithm) is proposed to measure the cluster quality with a minimum clustering error rate and fuzzy logic-based optimal cluster head selection to improve the stability and performance of the sensor network. The routing and clustering in WSNs are proposed in [10] by using the Grey Wolf Optimization technique with two innovative fitness solutions which indicate the better performances pertaining to clustering and routing in comparison with the existing algorithm. The opportunities for fog computing in WSNs are described in [11], where simulations are carried out to explore the design parameters. The overall performances of the sensor network are increased which uses the genetic algorithm in association with virtual grid-based dynamic route adjustment (VGDRA) in [12]; a comparative discussion with LEACH is also presented. A classification technique based on GA is discussed in [13] to calculate the multiple positions in a 3-D space using the direction-of-arrival method. A hybrid GA is implemented in a heterogeneous sensor network using BMHGA (greedy initialization and bidirectional mutation operations) for achieving the full coverage area for monitoring [14]. The optimization solution in WSNs is proposed in [15] using multi-objective wireless network optimization using GA (MOONGA) for the optimal deployment of the nodes in terms of different topology, applications, environment constraints, etc. A framework is provided in [16] to optimize sensor nodes dynamically by implementing genetic algorithm-based self-organizing network clustering (GASONeC). In this, the residual energy, distance, and numbers of nodes are optimized dynamically to ensure better performance in all conditions.

Similarly, ant colony optimization (ACO) finds many applications in the field of WSN. Some of the kinds of literature are discussed in [17–27]. The deployment of nodes in a sensor node is addressed by using minimum cost reliability constrained sensor node deployment problem (MCRC-SDP) in [17] to minimize the cost where ant colony optimization is also coupled for better performance in different environmental conditions. An improved ACO is proposed in [18] to enhance the life cycle with minimal energy consumption in a sensor network. Here, the message delivery is achieved with high accuracy and safety with the raise of speed in data packet communication. The EBAR (energy-efficient load-balanced ant-based routing) algorithm is implemented in [9] for optimizing the route establishment with reducing energy consumption where simulation results revealed the improvement in the performance in comparison with the traditional AC algorithm. A novel routing technique is proposed in [20] which is based on ACO to manage the network resources in real-time conditions. This method also prolongs the network life by choosing an optimal path with regard to low energy consumption. In [21], a secure routing protocol based on multi-objective ACO (SRPMA) is implemented in the sensor network by considering residual node energy and trust value of route path as two basic objectives. This also ensures protection from black hole attack during routing. A combination of AC-based algorithm with greedy migration mechanism is discussed in [22] to ensure full coverage with minimum deployment cost. It also confirms the GCLC (guaranty connectivity and low cost) with a steady power node to boost the network lifetime. ACO is also discussed in [23–25] to ensure the shortest path with minimum energy consumption in the sensor nodes along with the comparative report with the existing algorithm. A review is purposed regarding the different research based upon the ACO in [26], and a modified ACO is also purposed to solve the routing problem in the sensor network. For the selection of optimal cluster head among a group of nodes, the butterfly optimization algorithm (BOA) is employed in [27], where a comparison study with traditional approaches is also discussed.

The PSO (particle swarm optimization) technique is also found in a wide variety of applications in the WSN field, and some of the works of literature are discussed in [28–35]. A survey regarding the routing protocols pertaining to their structure, complexity, energy efficiency, and path establishment of a sensor network is thoroughly discussed along with a comparative report between classical and swarm-based protocols in [28]. The PSO algorithm is implemented in [29] to find the solution regarding the coverage problem in a sensor network. Here, the flight speed is properly mentioned to confirm the full coverage in the network. The PSO with a mobile sink is proposed in [30] to achieve the normal efficiency and quality of performance. Data security is also achieved in this work along with the enrichment of node lifetime. An accurate localization algorithm by using PSO is applied in [31] to improve the path-finding strategies in WSN, and it found much better performance in an unknown environment. Taylor C-SSA (Cat Salp Swarm Algorithm) is presented in [32] to address the energy problem in the network which is nothing but an energy-efficient multi-hop routing. To improve the node lifetime, the PSO is implemented in [33] which focuses on the cluster formation and the selection of cluster head. Also, the study is compared with the existing LEACH algorithm and found better performance. The artificial bee colony optimization (ABCO) which is based on PSO with the combination of LEACH algorithm is implemented in [34] to test the WSN performance in the diversified scenario. The PSO technique is realized in [35] for optimal selection of cluster head which reduces the price for tracing the optical spot of the cluster head node.

Aside from these three algorithms, some other algorithms are very popular in the field of WSN which are discussed in [27,36,37]. The Adelson-Velskii and Landis (AVL) tree rotation clustering algorithm is discussed in [36] to enhance the network lifetime by minimizing the delay and optimizing the energy efficiency of a sensor network. The enhancement of network lifetime is also discussed in [27] by using the butterfly optimization algorithm where cluster head selection is made by considering several factors like residual energy of nodes, distance to the neighbor nodes and base station, and centrality. The performance of this study is also compared with some relative techniques like low energy adaptive clustering hierarchy (LEACH), distributed energy-efficient clustering (DEEC), clustering and routing in WSNs using harmony search (CRHS), biogeography-based energy-saving routing architecture (BERA), cloud particle swarm optimization (CPSO), fractional lion optimization (FLION), etc. Here, the combination of BOA and CA is employed to choose the optimal cluster head and an optimal route respectively. Cluster head selection is also proposed in [37] of a sensor network using quasi-oppositional BOA along with the comparative study with original BOA.

1.3   GENETIC ALGORITHM (GA)

The genetic algorithm is one of the first and most well-regarded evolutionary algorithms, which mimics one of the most fundamental and well-known theories in the field of biology called the Darwinian theory of evolution proposed by Charles Darwin. It was developed by John Henry Holland in the 1970s. The main inspiration of the genetic algorithm is natural selection, i.e., the survival of the fittest, one of the key mechanisms in the theory of evolution. The selection is based on the variations in the genotype of an organism that increase its chances of survival. These genetic variations are preserved in nature and transferred from generation to generation. Good genes with desirable traits are produced by recombination or crossover of chromosomes and are passed to the next generation through inheritance. However, crossover and natural selection are not sufficient enough to increase genetic variations. Therefore, mutation plays a key role in evolution by adding new features to the population and promoting the diversity of the generation. Mutations are the random changes in a genetic sequence that occur in chromosomes during the process of recombination. However, it may result in the development of either good traits or bad traits in a population. Mutated genes with better features are preserved and transferred to the next generation through natural selection, while those features that result in less fit organisms are eliminated by nature through recombination and natural selection. The mutation also helps in recovering the traits that might disappear in subsequent generations due to crossover and constant competition between organisms. Hence, the main objective of all the evolutionary mechanisms is to simply maximize the survival of the population through each generation.

In order to apply the genetic algorithm to an optimization problem, two main components are required: (1) genetic representation of candidate solutions, and (2) fitness function to evaluate the solutions. In this algorithm, the standard representation of a candidate solution is an array of bits, also known as bitmap or bit string, as shown in Figure 1.1.

The size of the bit string is fixed unless the number of parameters of the optimization problem changes. Other types and structures of arrays can also be used instead of bit strings, for example, the array of integers, floating-point numbers, characters and operators, and even the array of pieces of codes to optimize a program (Genetic Programming). A fitness function, also called cost function or objective function, is used to evaluate a gene representation for determining how good the given combination of bits is with respect to the problem in consideration. As seen in Figure 1.2, it takes bit string as input and generates a unique fitness value as the output. The fitness function could be a mathematical equation, a computer program, or even a simulator.

In the genetic algorithm, a population refers to a set of candidate solutions, also known as chromosomes, which consists of a set of genes represented in the form of

0	1	1	0	1	0	1	0	0	1

FIGURE 1.1   Bit string representation of a chromosome in the genetic algorithm.

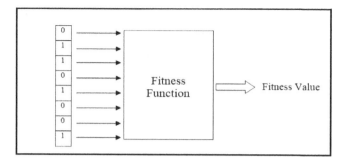

FIGURE 1.2   Fitness function in the genetic algorithm.

the bit string. Each bit value or gene represents a value for each variable of the given optimization problem. The initial generation of solutions is completely random and is subjected to changes by recombination, mutation, and selection in every generation, to be able to maximize or minimize a fitness function.

1.3.1   ROULETTE WHEEL

The roulette wheel is a stochastic operator which simulates the natural selection of the fittest individual. It is implemented by normalizing the fitness values of all the individuals, sorting those values in descending order, calculating the cumulative sum of each of these values, and finally visualizing the intervals between these cumulative sums as blocks on the number line. If we wrap these blocks around a circle, we will obtain a roulette wheel, in which each block represents the probability of a particular individual getting selected through natural selection. A block is selected in the roulette wheel when a random number generated in the interval [0, 1] lies within the interval of that block. The fittest individuals having a cumulative sum equal to or close to 1 are more likely to be selected since they occupy larger segments in the roulette wheel.

1.3.2   PROBABILITY OF CROSSOVER (P_c)

The genetic algorithm selects two fittest individuals as parents using the roulette wheel to perform recombination. In every generation, the number of chromosomes in the population should remain fixed. This can be done using single-point crossover, in which the genes of two parents are swapped before and after a particular point which is chosen randomly in every generation. However, all the chromosomes do not necessarily undergo crossover since, in nature, few chromosomes might survive more than one generation, and they still have the chance to be recombined with other chromosomes. The genetic algorithm simulates this by using a stochastic component called the probability of crossover, which lies in the range [0, 1]. To decide which individual goes to the next generation, a random number is generated in the interval [0, 1]. Two random numbers, R_1 and R_2, are generated for creating two children, C_1 and C_2, from parents P_1 and P_2 by crossover, which results in four different cases;

a. If $R_1 \leq P_c$ and $R_2 \leq P_c$, both C_1 and C_2 are transferred to the next generation.
b. If $R_1 > P_c$ and $R_2 > P_c$, both P_1 and P_2 are transferred to the next generation.
c. If $R_1 > P_c$ and $R_2 \leq P_c$, P_1 and C_2 are transferred to the next generation.
d. If $R_1 \leq P_c$ and $R_2 > P_c$, C_1 and P_2 are transferred to the next generation.

This is how chromosomes or candidate solutions are transferred from one generation to the next generation using the probability of crossover.

1.3.3 PROBABILITY OF MUTATION (P_m)

The probability of mutation is a predefined number that is chosen between the interval $[0, 1]$. Some of the genes face random changes regardless of their current fitness value. To mutate a gene, a random number is generated between the interval $[0, 1]$. If it is greater than or equal to P_m, the gene is mutated; otherwise, it is left unchanged.

1.3.4 ELITISM

Elitism is a technique that is used to copy and save a small portion of the best individuals called elites in every generation without changing the next generation of the population. It is done to avoid the loss of good traits after a certain number of generations due to the recombination and mutation operators. Instead of wasting time and computational resources in recovering lost features, a portion of the best solutions are reserved based on the elitism ratio ($E_r \in [0, 1]$) and transferred unchanged to the next generation. The elites can then be used in the process of selection, recombination, and mutation and help to improve the fitness value of solutions in the subsequent generations.

1.4 ANT COLONY OPTIMIZATION (ACO)

The first version of the ACO algorithm known as the Ant System, was developed by Marco Dorigo in 1992. This algorithm was inspired by the 'stigmergy' in nature, which refers to the communication and coordination among organisms that develop complex and decentralized intelligence without planning and direct interaction; for example, the foraging behavior of ants in nature. Ants produce chemicals called pheromones to communicate and find the shortest route from their nest to the food source. Pheromones are used to mark the path towards the food source since most of the ants are blind and hence are more likely to select a path with higher pheromone concentration. Initially, there could be multiple paths deposited with pheromones by multiple ants, but towards the end, only one path will be established between the nest and the food source with higher pheromone concentration, and all other paths with less pheromone concentration will be removed due to evaporation. A high pheromone level in a path signifies that it is the shortest path that attracts more ants towards it. A shorter path is deposited with pheromone faster than the longer path, and it will also have more pheromone concentration than the longer path due to evaporation. Therefore, the probability of selecting the shortest path increases every time the ants' commute, which makes them dominate other paths. With this simple technique, ant colonies are always able to find the shortest route between their nest and the food source.

1.4.1 MATHEMATICAL MODEL OF ACO

The paths chosen by the ants are represented by edges on a graph, as shown in Figure 1.3. Each edge connects two nodes (i^{th} node j^{th} and node, for example). The amount of pheromone deposited by each ant on an edge is represented as:

$$\Delta\tau_{i,j}^{k} = \begin{cases} \dfrac{C}{L_k}, \text{if } k^{th} \text{ ant travels on edge } ij \text{ in the graph} \\ 0, \text{if no ant travels the edge } ij \text{ in the graph} \end{cases}$$

where, L_k is the length of the path traversed by k^{th} ant and C is a constant.

The pheromone level $\Delta\tau_{i,j}^{k}$ is inversely proportional to the length of path L_k.

The amount of pheromone deposited by the total number of ants on each edge (considering vaporization) can be calculated as;

$$\tau_{i,j}^{k} = (1-\rho)\tau_{i,j} + \sum_{k=1}^{m}\Delta\tau_{i,j}^{k}$$

where,

$$\sum_{k=1}^{m}\Delta\tau_{i,j}^{k} \rightarrow \text{is the amount of new pheromone deposits}(\text{without vaporization})$$

$$m \rightarrow \text{total no.of ants}$$

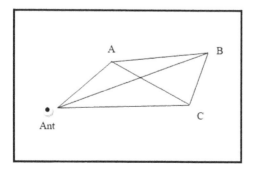

FIGURE 1.3 The path traversed by the Ant in the ant colony optimization algorithm.

$$\tau_{i,j} \rightarrow \text{current pheromone levels}$$

$$\rho \rightarrow \text{evaporation rate}$$

Evaporation is at the maximum level when $\rho = 1$, which means all the pheromones are removed as soon as they get deposited on the edge.

To implement this mathematical model and visualize the graphs, we need two matrices, one for the cost of an edge and the other for the pheromone concentration. These matrices are updated in each iteration until the shortest path is established. Ants choose a path using probabilities which can be calculated using the pheromone matrix as:

$$P_{i,j} = \frac{\left(\tau_{i,j}\right)^{\alpha} \left(\eta_{i,j}\right)^{\beta}}{\sum\left(\left(\tau_{i,j}\right)^{\alpha} \left(\eta_{i,j}\right)^{\beta}\right)}$$

where,

$$P_{i,j} \rightarrow \text{the probability of choosing the edge } ij \text{ on the graph}$$

$$\eta_{i,j} = \frac{1}{L_{i,j}} \rightarrow \text{the quantity of edge } ij$$

$$\tau_{i,j} \rightarrow \text{pheromone level}$$

The initial pheromone concentrations are identical for all the edges so that their impact on calculating the probabilities is similar. Tuning the parameters α and β, we can increase or decrease the impact of τ or η in the process of decision making. Out of all different edges starting from a particular node, the edge with a higher value of probability is chosen by the majority of ants. This is how the ACO algorithm helps in selecting the nodes for finding the optimum path in a routing problem.

1.5 PARTICLE SWARM OPTIMIZATION ALGORITHM (PSO)

The PSO algorithm is a population-based stochastic algorithm that mimics the act of finding direction and foraging of a flock of birds or school of fishes. It was developed by Kennedy and Eberhart in 1995. In this algorithm, the candidate solutions called particles start with a random direction in the problem space and follow a search strategy, which basically involves three steps, in order to find the optimal solution. The first step is to find the person's best location in the search landscape visited so far. For example, in a minimization problem, the personal best location would be the deepest point in the search landscape of its current iteration. The second step is to locate the group's best location in the landscape found by the group of particles in that particular iteration. The third step is to identify the particle's current travel direction to trace

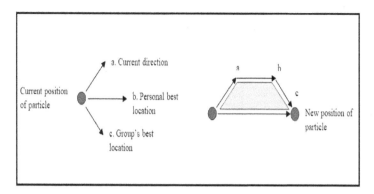

FIGURE 1.4   Search strategy of particles in particle swarm optimization.

the next location. In every iteration, each particle traverses a certain distance towards the current movement direction, the personal best location, and the swarm's best location to get to a new location in the landscape. An illustration of this traversal is shown in Figure 1.4. This distance is determined by a random component 'r' in the range of [0, 1]. The new location will be the starting point of the route for the next iteration.

If the particle's new position is better than its personal best location, the particle's record is updated. Thus in every iteration, the new location of the particle is updated and the search area (gray shaded region in Figure 1.1) changes depending on this location. As the group maintains the best locations in the region and searches around them only, the probability of finding the global optimal solution is high. To find the global optimum, we need to slow down the search process by reducing the distance covered by the particle proportional to the number of iterations, which in turn results in the search area becoming smaller and the search process becoming local rather than global in every iteration. This is how the PSO algorithm performs search operations by greatly reducing the time complexity of finding the best solution.

1.5.1 Mathematical Model of PSO

In order to update the position of each particle in the search landscape, two vectors are considered, (i) a position vector and (ii) a velocity vector. The position vector of the particle is represented as $\overrightarrow{X_i^t} = \begin{bmatrix} x_i^t, y_i^t, z_i^t \end{bmatrix}$, where $\overrightarrow{X_i^t}$ is a three-dimensional vector, 'i' in the subscript corresponds to i^{th} particle and 't' is the iteration number. $\overrightarrow{X_i^t}$ can also be an n-dimensional vector, written as $\overrightarrow{X_i^t} = \begin{bmatrix} x_i^t, y_i^t, z_i^t, \ldots, n_i^t \end{bmatrix}$. To update the position vector in every iteration, we need to define the direction and intensity of the movement using the velocity vector. It is represented as:

$$\overrightarrow{V_i^{t+1}} = \omega \overrightarrow{V_i^t} + c_1 r_1 \left(\overrightarrow{P_i^t} - \overrightarrow{X_i^t} \right) + c_2 r_2 \left(\overrightarrow{G^t} - \overrightarrow{X_i^t} \right) \qquad (1.1)$$

where,

$\overrightarrow{V_i^{t+1}} \rightarrow$ velocity of in $(t+1)^{th}$ iteration

$\overrightarrow{V_i^t} \rightarrow$ velocity in t^{th} iteration

$\omega\overrightarrow{V_i^t} \rightarrow$ inertia

$\omega \rightarrow$ inertia weight

$c_1 r_1 \left(\overrightarrow{P_i^t} - \overrightarrow{X_i^t} \right) \rightarrow$ cognitive component or individual component

$\left(\overrightarrow{P_i^t} - \overrightarrow{X_i^t} \right) \rightarrow$ distance to the personal best solution

$\overrightarrow{P_i^t} \rightarrow$ personal best solution

$c_2 r_2 \left(\overrightarrow{G^t} - \overrightarrow{X_i^t} \right) \rightarrow$ social component

$\left(\overrightarrow{G^t} - \overrightarrow{X_i^t} \right) \rightarrow$ distance to the global best solution

$\overrightarrow{G^t} \rightarrow$ global best solution

$r_1, r_2 \rightarrow$ random components in the range of $[0, 1]$

Therefore, the new location of the particle after every iteration can be represented by its position vector, which can be calculated as:

$$\overrightarrow{X_i^{t+1}} = \overrightarrow{X_i^t} + \overrightarrow{V_i^{t+1}} \tag{1.2}$$

where,

$\overrightarrow{X_i^{t+1}} \rightarrow$ position of the particle in $(t+1)^{th}$ iteration

$\overrightarrow{X_i^t} \rightarrow$ position in t^{th} iteration

$\overrightarrow{V_i^{t+1}} \rightarrow$ Velocity in $(t+1)^{th}$ iteration

Both the velocity vector and position vector are updated in each iteration with Equations (1.1) and (1.2) respectively. The social component determines the distance between a particle's current location and the best location found by the entire swarm. The impact of cognitive component and social component on the movement of particles can be varied by tuning the coefficients c_1 and c_2. The exploration and exploitation of the search space are tuned using the inertia weight parameter. Exploration is at the minimum level when $\omega = 0$, whereas exploitation is the maximum level when $\omega = 1$. It is necessary to balance both exploration and exploitation using the inertia parameter during the optimization process.

1.6 IMPLEMENTATION OF GENETIC ALGORITHM IN A WSN

The deployment of a large number of sensor nodes in a WSN necessitates the use of proper routing techniques in order to extend the lifespan of WSNs and minimize energy consumption. However, routing of data packets from source to destination turns into a very complicated and challenging task when the network is huge with a large number of nodes. Many shortest path algorithms like Dijkstra's algorithm face various drawbacks while solving the routing problem due to a large number of calculations in every iteration. This highly increases the time complexity and the usage of computational resources in finding the optimal path. In such cases, the genetic algorithm can be used to find the best route which can balance energy and data load in a network. The genetic algorithms are useful for non-deterministic polynomial-time hardness (NP-hard) problems and can be implemented to find solutions to the optimization problems like the Traveling Salesman Problem in very little time complexity. We can use this algorithm in connecting a set of sensor nodes to achieve the shortest route for traversing the complete sensor network while visiting every sensor node exactly once.

In the proposed algorithm, the sensor nodes are represented in the form of xy-coordinates on a graph. These coordinates are represented in a matrix as:

$$xy\_coord = \begin{bmatrix} 172\ 426;\ 183\ 863;\ 625\ 285;\ 63\ 295;\ 29\ 86;\ 44\ 336;\ 93\ 478;\ 361\ 12;\ 539\ 594; \\ 189\ 261;\ 961\ 411;\ 635\ 80;\ 535\ 420;\ 467\ 145;\ 40\ 243;\ 221\ 569;\ 223\ 318;\ 727\ 997 \end{bmatrix}$$

To implement the proposed algorithm, first, a set of random populations is generated and initialized. The solutions are constantly modified by recombination, mutation, and selection in every iteration to maximize the fitness function. The fitness value for each individual is calculated after every iteration. The individual with a larger fitness value results in a better solution. The total distance and the least cost are calculated based on these solutions and are updated in every iteration until the best cost is found and the shortest path is established.

1.6.1 SIMULATION RESULTS

In the proposed algorithm, the size of the population is set to 400, and the number of generations is 80. The optimal route thus established by the genetic algorithm is as follows –

$$optimal\_route = 13\ 3\ 11\ 12\ 14\ 8\ 17\ 10\ 15\ 4\ 6\ 7\ 1\ 16\ 5\ 2\ 18\ 9$$

The length of this route calculated by the algorithm is;

$$min\_dist = 4.0657e + 03$$

The optimized shortest path of the network connecting all the sensor nodes is shown in Figure 1.5. In this figure, the purple points represent the sensor nodes, and the line passing through all the points is the path established by the genetic algorithm.

The total number of iterations required to find the optimum value is: current_gen = 51.

The variation of objective function value with iteration count is shown in Figure 1.6. As seen in this figure, the length of the path connecting all the nodes decreases with every generation, and the algorithm obtains the shortest path at iteration number 51.

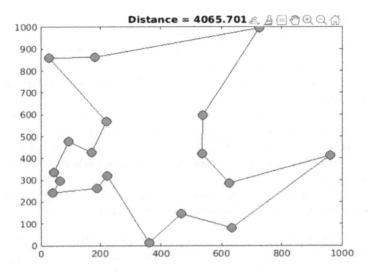

FIGURE 1.5   Shortest path traversing the sensor nodes as per the genetic algorithm.

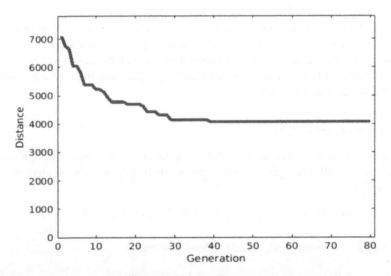

FIGURE 1.6   Variation of path length with each generation as obtained by the genetic algorithm.

1.7   IMPLEMENTATION OF ANT COLONY OPTIMIZATION ALGORITHM IN A WSN

Another technique that can be used in routing operations of WSNs is the ACO approach. Since WSNs consist of a large number of nodes with a limited power supply to collect and transmit useful information from source to destination, it is important to gather the information in an energy-efficient manner. The ACO algorithm is a highly adaptive and robust method for network routing. It can be implemented in optimization problems like the Traveling Salesman Problem to find solutions in less time complexity. Like the genetic algorithm, the ACO algorithm can also be used in the connecting set of sensor nodes for achieving the shortest route for traversing the network while visiting every sensor node exactly once.

In the proposed algorithm, the same set of xy-coordinates representing the sensor nodes on a graph is used in the form of a matrix;

$$xy\_coord = \begin{bmatrix} 172\ 426; 183\ 863; 625\ 285; 63\ 295; 29\ 86; 44\ 336; 93\ 478;\ 361\ 12;\ 539\ 594; \\ 189\ 261; 961\ 411; 635\ 80; 535\ 420; 467\ 145; 40\ 243; 221\ 569;\ 223\ 318;\ 727\ 997 \end{bmatrix}$$

To implement the proposed algorithm, first, a graph is created using the given xy-coordinates, and edges are calculated and drawn from one node to another using the Euclidean distance between nodes, as shown in Figure 1.7. In this figure, the purple points represent the sensor nodes, and the lines connecting these points are the edges that mark several possible paths which can be traversed in the algorithm.

Then, a colony of artificial ants is created in which each ant is placed on a randomly chosen initial node. A pheromone matrix is initialized and updated once all the ants have constructed a tour. Each ant constructs a tour on one of the edges probabilistically based on the pheromone levels on the edges. The current partial tour of each ant is stored in a matrix to determine the set of nodes visited. The pheromone levels on the visited edges are lowered in every iteration by the evaporation factor so that the ants can avoid taking longer routes with less pheromone concentration. This improves the tours constructed by the ants and thus improves their fitness value. In the final iteration, the fitness value of all the ants is calculated to obtain the best ant, and hence, the route established by the best ant is the shortest path.

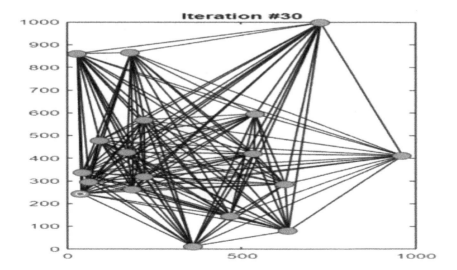

FIGURE 1.7 Edges connecting the sensor nodes that represent possible paths for traversal.

1.7.1 Simulation Result

In the proposed algorithm, the initial parameters of ACO are assigned the following values:

Total number of iterations = 30
Total number of ants = 800
Evaporation rate = 0.5

The pheromone concentration on all the edges at the end of the last iteration can be visualized in Figure 1.8. In this figure, the paths that are darker in color and have more width are the paths that are traversed a greater number of times. These paths are having higher pheromone concentration in the graph.

The value of the objective function—that is, length of the shortest path obtained in every iteration by the ACO algorithm—is shown in Figure 1.9. This length decreases with each subsequent iteration to result in the shortest path connecting all the sensor nodes in the graph.

Therefore, the length of the optimum path found by the ACO algorithm is 4065.7014 units. The shortest route connecting all the sensor nodes can be visualized in Figure 1.10, in which the purple points represent the nodes and the line passing through all the points is the shortest path established by the ACO algorithm.

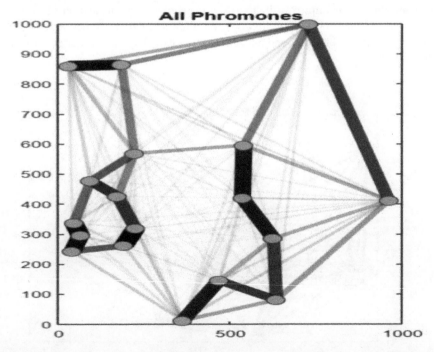

FIGURE 1.8 Pheromone concentration on the edges of graph marked by the ACO algorithm.

```
Iteration #1 Shortest length = 5395.3349
Iteration #2 Shortest length = 4555.7177
Iteration #3 Shortest length = 4436.9199
Iteration #4 Shortest length = 4419.1369
Iteration #5 Shortest length = 4368.3582
Iteration #6 Shortest length = 4368.3582
Iteration #7 Shortest length = 4294.4929
Iteration #8 Shortest length = 4219.8769
Iteration #9 Shortest length = 4190.3748
Iteration #10 Shortest length = 4084.1002
Iteration #11 Shortest length = 4084.1002
Iteration #12 Shortest length = 4084.1002
Iteration #13 Shortest length = 4084.1002
Iteration #14 Shortest length = 4084.1002
Iteration #15 Shortest length = 4084.1002
Iteration #16 Shortest length = 4084.1002
Iteration #17 Shortest length = 4084.1002
Iteration #18 Shortest length = 4084.1002
Iteration #19 Shortest length = 4084.1002
Iteration #20 Shortest length = 4084.1002
Iteration #21 Shortest length = 4084.1002
Iteration #22 Shortest length = 4084.1002
Iteration #23 Shortest length = 4084.1002
Iteration #24 Shortest length = 4084.1002
Iteration #25 Shortest length = 4084.1002
Iteration #26 Shortest length = 4065.7014
Iteration #27 Shortest length = 4065.7014
Iteration #28 Shortest length = 4065.7014
Iteration #29 Shortest length = 4065.7014
Iteration #30 Shortest length = 4065.7014
```

FIGURE 1.9   Variation of objective function value with each iteration in the ACO algorithm.

1.8   IMPLEMENTATION OF PARTICLE SWARM OPTIMIZATION ALGORITHM IN A WSN

In WSNs, the maximization of network lifetime is one of the major challenges of the communication network, since the sensor nodes have a limited energy supply. Hence, proper localization of nodes and clustering is necessary to expand the network lifespan and minimize the consumption of power. The PSO algorithm is used in a WSN to optimize the LEACH protocol (low-energy adaptive clustering hierarchy) by identifying the cluster head, which aggregates and processes the data transmitted by the cluster member nodes and forwards it to the base station, in order to minimize the overall energy consumption of the sensor nodes. The optimal selection of the cluster head is based on the residual energy of the sensor nodes and the cost of communication from these nodes to the sink node (base station).

In the clustered sensor network, the nodes are assumed to be stationary. The goal is to find an optimal location for the cluster head, and it should be close to the center of

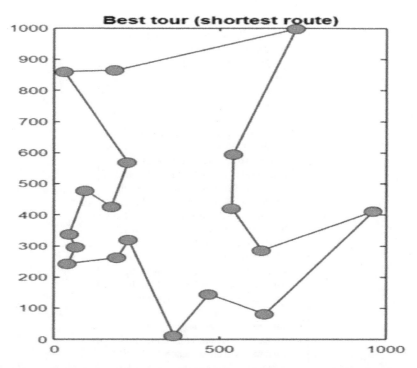

FIGURE 1.10   Shortest path connecting all the nodes obtained by the ACO algorithm.

mass (COM) of the cluster which is the midpoint of the sensor node distribution within the cluster. As shown in Figure 1.11, the COM is located at (20, 20) in the graph.

In the proposed technique, PSO considers a swarm of eight particles or random solutions that slowly converge towards the COM with every iteration for finding the best location of the cluster head. Such localization of the cluster head would

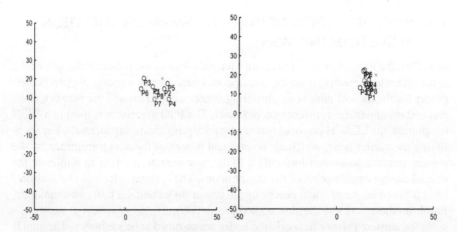

FIGURE 1.11   Center of the mass of the cluster of sensor nodes.

ultimately help in minimizing the average distance covered by the sensors in transmitting data to the cluster head.

To implement the proposed PSO-based clustering algorithm, a set of variables and PSO parameters are added and initialized with certain values. These parameters include the maximum number of iterations, swarm size, inertia weight, coefficient of personal acceleration, and social acceleration. The variables include the position and velocity of particles, the personal best solution, the global best solution, the energy of the sensor nodes, and cost value. The best values for these variables are obtained by running the algorithm through multiple iterations. The energy lost due to the transmission of information is calculated and subtracted from the residual energy of nodes in every round of operation. Finally, the cost or fitness value of every particle is calculated using the cost function, which is based on the residual energy of the member nodes and the minimum average distance from these nodes. At last, the particle having the least cost is updated as the cluster head, which then forwards data to the sink node, optimizing the total energy.

1.8.1 Simulation Result

In the proposed algorithm, the PSO parameters are assigned the following values:

Maximum Number of Iterations = 25
Swarm Size = 8
Inertia Coefficient (w) = 1
Personal Acceleration Coefficient (c1) = 1.8;
Social Acceleration Coefficient (c2) = 1.8;

The value of the least cost and the corresponding particle associated with it in every iteration is shown in Figure 1.12. As seen in the figure, the value of the cost function decreases in every iteration as the total energy of the cluster also decreases in each round of iteration.

Therefore, particle number 2 is the last location of the cluster head in this case, with the least cost of 0.74455. The variation of total energy of cluster with iteration count is shown in Figure 1.13, in which the total residual energy keeps decreasing till iteration number 22.

1.9 COMPARATIVE ANALYSIS

It is evident from this study that the framework of all three bio-inspired algorithms is similar; that is, they begin with a population of random solutions, perform exploration and exploitation with specific operators, and then estimate the global optimum for a given optimization problem. However, different algorithms perform in different ways, giving different results for the same problem. This is due to the No Free Lunch (NLF) theorem, which states that no optimization algorithm can solve all the optimization problems. As a result, one algorithm can be very effective in solving a certain set of problems and may not be as effective on a different set of problems. In the study presented here, it is observed that the time complexity of ACO is more than that

```
Iteration 1: Best Cost = 9.0593: Particle =8
Iteration 2: Best Cost = 7.5382: Particle =6
Iteration 3: Best Cost = 7.5177: Particle =6
Iteration 4: Best Cost = 5.8761: Particle =2
Iteration 5: Best Cost = 4.7987: Particle =7
Iteration 6: Best Cost = 4.3165: Particle =2
Iteration 7: Best Cost = 3.3975: Particle =3
Iteration 8: Best Cost = 3.3975: Particle =3
Iteration 9: Best Cost = 3.3975: Particle =3
Iteration 10: Best Cost = 3.3975: Particle =3
Iteration 11: Best Cost = 2.9934: Particle =1
Iteration 12: Best Cost = 2.9934: Particle =1
Iteration 13: Best Cost = 2.9934: Particle =1
Iteration 14: Best Cost = 2.9143: Particle =2
Iteration 15: Best Cost = 2.8212: Particle =4
Iteration 16: Best Cost = 2.6989: Particle =5
Iteration 17: Best Cost = 2.4288: Particle =8
Iteration 18: Best Cost = 1.5595: Particle =7
Iteration 19: Best Cost = 1.5595: Particle =7
Iteration 20: Best Cost = 1.5595: Particle =7
Iteration 21: Best Cost = 0.98953: Particle =4
Iteration 22: Best Cost = 0.74455: Particle =2
Iteration 23: Best Cost = 0.74455: Particle =2
Iteration 24: Best Cost = 0.74455: Particle =2
Iteration 25: Best Cost = 0.74455: Particle =2
```

FIGURE 1.12   Variation of cost function value with iteration count in PSO algorithm.

of the genetic algorithm. Also, the genetic algorithm is easy to implement and cost-efficient in terms of computational resources. However, ACO can give better results with large problem size, whereas the genetic algorithm might not find the most optimal solution to a particular problem. In the case of PSO, the algorithm is easy to implement and has fewer parameters to adjust; however, it is iterative, and high usage of computational resources can prohibit its use in high-speed applications. The parameters of each algorithm depend on the nature of a problem, and depending on such problem, one technique may perform better than the other.

1.10   CONCLUSION AND FUTURE WORK

In this study, the significance of bio-inspired algorithms for the intelligent optimization of non-biological systems is elaborated with a comprehensive overview of three population-based algorithms, namely the genetic algorithm, ant colony optimization, and particle swarm optimization. To create these evolutionary and swarm-based algorithms, the key components of corresponding biological systems are observed and suitably modified so that they can fit the required use case.

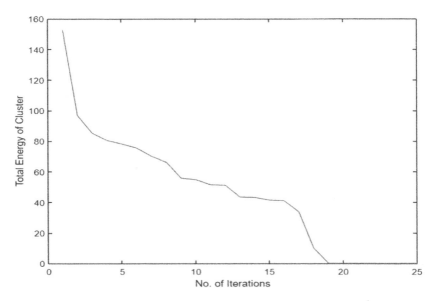

FIGURE 1.13   Variation of total energy of cluster with iterations as obtained by the PSO algorithm.

We have implemented these techniques in two of the most challenging issues in applications of wireless sensor networks, that is, the shortest path routing of data packets and energy efficient selection of cluster head. Through this, we were able to determine that bio-inspired algorithms are able to produce high-quality, optimal solutions that converge faster and have very low computational complexity. But they come with their shortcomings as well, since they tend to be resource-intensive as they consume a lot of memory and cannot be used in some real-time applications which demand fast processing. Based on this, future work needs to be done on minimizing the energy consumption on a node level as well as creating movement strategies that will not only help heal the network coverage in case of holes, but also improve the network lifetime.

The development of efficient optimization algorithms is the key to improving consumption of the limited resources of WSNs. Through our work, a comparative analysis of these algorithms is presented along with their advantages and limitations, which can be used as a future guide for the implementation of such algorithms in WSNs.

REFERENCES

1. Cosmena Mahapatra, Ashish Payal, Meenu Chopra, *"Review of WSN and its quality of service parameters using nature-inspired algorithm,"* International Conference on Innovative Computing and Communications, Part of the Advances in Intelligent Systems and Computing book series, 1059 (2019), Springer, Singapore, 451–461.
2. Eman Saad, Mostafa Elhosseini, Amira Yassin Haikal, "Recent achievements in sensor localization algorithms," *Alexandria Engineering Journal,* 57 (2018), 4219–4228.

3. Neha Sharma, Vishal Gupta, "Meta-heuristic based optimization of WSNs localisation problem – a survey," *International Conference on Smart Sustainable Intelligent Computing and Applications under ICITETM 2020, ScienceDirect, Procedia Computer Science*, 173 (2020), 36–45.

4. Hanaa ZainEldin, Mostafa A. Elhosseini, Hesham A. Ali, "Image compression algorithms in wireless multimedia sensor networks: A survey," *Ain Shams Engineering Journal*, 6 (2015), 481–490.

5. Shazana Md Zin, Nor Badrul Anuar, Laiha Mat Kiah, Ismail Ahmedy, "Routing protocol design for secure WSN: Review and open research issues," *Journal of Network and Computer Applications*, 41 (2014), 517–530.

6. Shazana Md Zin, Nor Badrul Anuar, Laiha Mat Kiah, Ismail Ahmedy, "Survey of secure multipath routing protocols for WSNs," *Journal of Network and Computer Applications*, 55 (2015), 123–153.

7. Waqas Rehan, Stefan Fischer, Maaz Rehan, Mubashir Husain Rehmani, "A comprehensive survey on multichannel routing in wireless sensor networks," *Journal of Network and Computer Applications*, 95 (2017), 1–25.

8. Gulshan Kumar, Mritunjay Kumar Rai, "An energy efficient and optimized load balanced localization method using CDS with one-hop neighbourhood and genetic algorithm in WSNs," *Journal of Network and Computer Applications*, 78 (2017), 73–82.

9. Amir Abbas Baradaran, Keivan Navi, "HQCA-WSN: High-quality clustering algorithm and optimal cluster head selection using fuzzy logic in wireless sensor networks," *Fuzzy Sets and Systems*, 389 (2020), 114–144.

10. Amruta Lipare, Damodar Reddy Edla, Venkatanareshbabu Kuppili, "Energy efficient load balancing approach for avoiding energy hole problem in WSN using Grey Wolf Optimizer with novel fitness function," *Applied Soft Computing*, 84 (2019), 105706.

11. Lelio Campanile, Marco Gribaudo, Mauro Iacono, Michele Mastroianni, "Performance evaluation of a fog WSN infrastructure for emergency management," *Simulation Modelling Practice and Theory*, 104 (2020), 102120.

12. Anurag Saini, Avnish Kansal, Navdeep Singh Randhawa, "Minimization of energy consumption in WSN using hybrid WECRA approach," *The 5th International Workshop on Wireless Technology Innovations in Smart Grid (WTISG), ScienceDirect, Procedia Computer Science*, 155,(2019), 803–808, August 19–21, 2019, Halifax, Canada.

13. Yuan Zhang, Yue Ivan Wu, "Multiple sources localization by the WSN using the direction-of-arrivals classified by the genetic algorithm," *IEEE Access*, 7 (2019), 173626–173635.

14. Jingjing Li, Zhipeng Luo, Jing Xiao, "A hybrid genetic algorithm with bidirectional mutation for maximizing lifetime of heterogeneous wireless sensor networks," *IEEE Access*, 8 (2020), 72261–72274.

15. S. E. Bouzid, Y. Seresstou, K. Raoof, M. N. Omri, M. Mbarki, C. Dridi, "MOONGA: Multi-objective optimization of wireless network approach based on genetic algorithm," *IEEE Access*, 8 (2020), 105793–105814.

16. Xiaohui Yuan, Mohamed Elhoseny, Hamdy K. El-Minir, Alaa M. Riad, "A genetic algorithm-based, dynamic clustering method towards improved WSN longevity," *Journal of Network and Systems Management*, 25 (2017), 21–46.

17. Dina S. Deif, Yasser Gadallah, "An ant colony optimization approach for the deployment of reliable wireless sensor networks," *IEEE Access*, 5 (2017), 10744–10756.

18. Kai-Chun Chu, Der-Juinn Horng, Kuo-Chi Chang, "Numerical optimization of the energy consumption for wireless sensor networks based on an improved ant colony algorithm," *IEEE Access*, 7 (2019), 2169–3536.

19. Xinlu Li, Brian Keegan, Fredrick Mtenzi, Thomas Weise, Ming Tan, "Energy-efficient load balancing ant based routing algorithm for wireless sensor networks," *IEEE Access*, 7 (2019), 113182–113196.
20. Amir Seyyedabbasi, Farzad Kiani, "MAP-ACO: An efficient protocol for multi-agent pathfinding in real-time WSN and decentralized IoT systems," *Microprocessors and Microsystems*, 79 (November 2020), 103325.
21. Ziwen Suna, Min Wei, Zhiwei Zhang, Gang Qu, "Secure routing protocol based on multi-objective ant-colony-optimization for wireless sensor networks," *Applied Soft Computing*, 77 (April 2019), 366–375.
22. Xuxun Liu, Desi He, "Ant colony optimization with greedy migration mechanism for node deployment in wireless sensor networks," *Journal of Network and Computer Applications*, 39 (March 2014), 310–318.
23. G. Gajalakshmi, G. Umarani Srikanth, *"A survey on the utilization of ant colony optimization (ACO) algorithm in WSN,"* 2016 *International Conference on Information Communication and Embedded Systems (ICICES)*, February 25–26, 2016, Chennai, India.
24. T.S. Nimisha, R. Ramalakshmi, *"Energy efficient connected dominating set construction using ant colony optimization technique in wireless sensor network,"* 2015 *International Conference on Innovations in Information, Embedded and Communication Systems (ICIIECS)*, March 19–20, 2015, Coimbatore, India.
25. Jian-Feng Yan, Yang Gao, Lu Yang, *"Ant colony optimization for wireless sensor networks routing,"* 2011 *International Conference on Machine Learning and Cybernetics*, July 10–13, 2011, Guilin, China.
26. B. Chandra, Mohan R. Baskaran, "A survey: Ant colony optimization based recent research and implementation on several engineering domain," *Expert Systems with Applications*, 39, 4, March 2012, 4618–4627.
27. Prachi Maheshwari, Ajay K. Sharma, Karan Verma, "Energy efficient cluster based routing protocol for WSN using butterfly optimization algorithm and ant colony optimization," *Ad Hoc Networks*, 110, 1 January 2021, 102317.
28. Adamu Murtal Zungerua, Li-Minn Ang, Kah PhooiSeng, "Classical and swarm intelligence based routing protocols for wireless sensor networks: A survey and comparison," *Journal of Network and Computer Applications*, 35, 5 (September 2012), 1508–1536.
29. Dianna Song, Jianhua Qu, *"A Fast efficient particle swarm optimization algorithm for coverage of wireless sensor network,"* 2017 *International Conference on Computer Systems, Electronics and Control (ICCSEC)*, December 25–27. 2017, Dalian, China.
30. K Vimal Kumar Stephen; V Mathivanan, *"An energy aware secure wireless network using particle swarm optimization"*, *Majan International Conference (MIC)*, March 19–20, 2018, Muscat, Oman.
31. Xihai Zhang, Tianjian Wang, Junlong Fang, *"A node localization approach using particle swarm optimization in wireless sensor networks,"* 2014 *International Conference on Identification, Information and Knowledge in the Internet of Things*, October 17–18, 2014, Beijing, China.
32. A. Vinitha, M.S.S. Rukmini, Dhirajsunehra, *"Secure and energy aware multi-hop routing protocol in WSN using Taylor-based hybrid optimization algorithm,"* *Journal of King Saud University—Computer and Information Sciences*, 2019.
33. Amita Yadav, Suresh Kumar, Singh Vijendra, *"Network life time analysis of WSNs using particle swarm optimization,"* *International Conference on Computational Intelligence and Data Science (ICCIDS 2018), ScienceDirect, Procedia Computer Science*, 132 (2018), 805–815.

34. Ankit Gambhira, Ashish Payal, Rajeev Arya, *"Performance analysis of artificial bee colony optimization based clustering protocol in various scenarios of WSN,"* International Conference on Computational Intelligence and Data Science (ICCIDS 2018), *ScienceDirect, Procedia Computer Science*, 132 (2018), 183–188.
35. Buddha Singha, D. K. Lobiyal, "Energy-aware cluster head selection using particle swarm optimization and analysis of packet retransmissions in WSN," C3IT, SciVerse ScienceDirect, *Procedia Technology*, 4 (2012), 171–176.
36. R. Raj Priyadarshini, N. Sivakumar, "Cluster head selection based on Minimum Connected Dominating Set and Bi-Partite inspired methodology for energy conservation in WSNs," *Journal of King Saud University—Computer and Information Sciences* (2018), 1–13.
37. Nageswara Rao Malisetti, Vinaya Kumar Pamula, *"Performance of quasi oppositional butterfly optimization algorithm for cluster head selection in WSNs,"* Third International Conference on Computing and Network Communications (CoCoNet-19), *ScienceDirect, Procedia Computer Science*, 171 (2020), 1953–1960.

2 An Improved Genetic Algorithm with Haar Lifting for Optimal Sensor Deployment in Target Covers Based Wireless Sensor Networks

T. Ganesan and Pothuraju Rajarajeswari

Koneru Lakshmaiah Education Foundation, Guntur, Andhra Pradesh, India

CONTENTS

2.1 INTRODUCTION

The wireless sensor network (WSN) is delivered spatially with abilities of sensing, wireless interaction, and data processing and is used in the monitoring environment. These sensor nodes combine to perform the very complicated sensing task in both accessible and inaccessible regions and data collection tasks in the same region [1,2]. WSNs are utilized for disaster prevention to target surveillance in the monitoring sector. The sensors depend on three different units, namely, communication range, sensing range, and energy of nodes [3,4]. The communication range cares about the responsibility of data transmission and data reception over the nodes. Energy has a responsibility for the efficient power supply of communication and sensing. Finally, sensing units sense the target object present in that region.

DOI: 10.1201/9781003145028-2

Since the limited energy parameter, the most important objective of most research is to maximize the network duration and energy optimization [5,6]. Thus, sensors are categorized into different sets, specifically sensor cover sets, and the limited number of sensor nodes use the increase in the network lifetime. WSNs can configure themselves automatically to convene the several application environments based on the requirements. Usually, in the study area of preserving target coverage, the main goal is to obtain complete coverage if possible, by a limited number of sensor node set groups [1,5].

The sensors may be distributed in one of two ways, random deployment or deterministic or planned deployment, depending on the application. Random sensor deployment is utilized when the sensing area is a larger, remote, hostile, and inaccessible region, such as a dense forest area, undersea, an eruption of volcanoes, or an active war zone [7–9]. In this case, random deployment might be the best choice, done by aircraft in the sky or in some other manner. In this situation, many sensors are deployed for redundant covers which reduce the network lifetime. There are a plethora of ways being done on how to control this density to some level [7,10].

Likewise, deterministic sensor placement in an accessible region is used to locate optimal locations of sensor nodes when network design goals can be fulfilled [11,12]. The network coverage, cost, lifetime, and node connectivity are the objectives of network design, which reduce sensor redundancy, computation complexity, and resource utilization [13,14]. The coverage can be categorized into area, target, and barrier coverage, which monitors the entire area, a set of specific points, and country border coverage, respectively. A thorough investigation has been undertaken in the topic of WSN deployment challenges in terms of target coverage [15,16].

This chapter's main contribution to covering the maximum amount of target points with a limited amount of sensor nodes, and every sensor node can be established a communication connection with other sensor nodes. First, the genetic algorithm (GA)-based random population matrix contains the sensor coordinate position. This matrix can be applied to crossover and mutation operation to improve the solution quality. Second, these sensors coordinate the population matrix that can be applied to a local enhancement of lifting wavelet transform for fast computation of optimization. The rest of the chapter is structured as follows. Section 2.2 presents the correlated works in WSNs considering the target coverage, node connectivity, and energy analysis. Section 2.3 introduces the design of the proposed method of problem formulation and quality measure of target coverage and node connectivity. Section 2.4 presents the 2D discrete Haar lifting wavelet transform, which adjusts the sensor coordinate position. Section 2.5 discusses the mathematical computation of GA with 2DDHLWT for the final sensor coordinate position. Section 2.6 analyzes the simulation works and results in discussions, and Section 2.7 concludes the work presented in the chapter.

2.2 RELATED WORKS

Several researchers proposed wavelet transform in the wireless sensor network to resolve several problems. As our recommended effort is a genetic algorithm with wavelet transform for sensor node positioning in an optimal position, we will review here the different existing algorithms in this regard.

The maximum number of sensor disjoint set is organized by the authors of [1], whereas to achieve a maximum lifetime of the sensor network using disjoint set covers (DSC) for covers and lifetime for dynamic coverage maintenance (DCM). The DSC problem was solved when each set of sensors were switched on and the DCM problem can be resolved using when the independent set extended the working periods. The hybrid memetic algorithm (MA) established organizing strategy explored the maximum number of DSC and then heuristic recursive algorithm (HRA) is used for patching the coverage hole while node failure and energy problems in nodes. The authors of [10] stated that the optimum locations of limited sensor nodes are able to satisfy the coverage requirements. Generally, Monte Carlo GA depicted the circular sensing range of sensor nodes. The sensing model of sensor nodes are defined in two ways: Euclidean distance between sensor and demand point in the two-dimensional region as a binary model, and the virtual force algorithm (VFA) based probability model. The demand point coverage probability is divided into fully covered, partially covered as considered uncover, cover, and uncovered pattern. The random location of the sensor node can be initialized in Harmony Search (HS). The quality of the solution is obtained in Harmony Memory (HM).

The authors of [11] pointed out the set k-cover problem using an integer coded memetic algorithm (MA). The integer can be represented by a tighter upper bound. MA randomly selects the smallest covered targets depending on the sensor as critical sensors [6]. Mostly, a recycling operator is used to reallocate the redundant sensors (non-critical sensors) to cover more targets. Recently, the authors of [5] showed an optimized k-coverage of mobile networks. A GA-based, random-generated initial population selected the optimum number of cluster heads. Each transmission round the sensor networks was adjusted for the next round. Also, for each round, the expected energy was computed for the next round as a fitness function.

The authors of [13] applied integer-coded GA for area coverage, which is based on Laplace crossover (LX) and Arithmetic Crossover methods (AXMO) operators and special local examination of virtual force algorithm (VFA). VFA targeted the maximum area coverage; meanwhile, it pulled all non-overlapped sensors. The authors of [9] considered a nonlinear optimized problem with binary variables for the sensor's distribution, connectivity, and reliability. The efficient heuristic GA investigated the uncertainty-aware, cluster-based sensor deployment that had formulated the multi-objective optimization problems. The network topology considered reduced cluster heads, orphans' sensors, orphans' cluster head, and overlapped area of clusters with unnecessary retransmission. The deployment point was labeled with two-bit string values; that is, "00" means sensors are not deployed, "01" or "10" means regular sensor deployment, and "11" is for cluster head deployment.

The authors of [15] formulated the integer linear programmed (ILP) target coverage, which is established on the exact algorithm to determine the optimal solution. The covering disk represents the radius in the target disk; if the target was inside the covering disk, then only the sensor can cover the target. The sensor may cover as many as targets, which was in an intersection region. Then the authors concentrated on the network connectivity, which is based on a constant-approximation algorithm.

Moreover, the authors of [2] named coverage and node placement as Minimum Perpetual Coverage Node Placement (MPCNP). The authors determined the locations for the quality of target coverage and sufficient energy of nodes. ILP was used to reduce the number of sensors for MPCNP. Before deploying the sensor, nodes are considered as Greedy Round (GRNP), Target Protection (TPNP), and Energy Efficient (EENP) node placements.

The authors of [16] have brought in the concepts of sensor deployment for smart cities. The geometric k-center wireless gateway distribution can be solved using modified particle swarm optimization. It determined the distribution location and amount of the access point based on the regional facts. The k-coverage and m-connectivity sensor deployment was assessed [17], with immigrant imperialist competitive algorithm (IICA)-based node placements. The authors of [7] presented the deterministic sensor placement using a hybrid memetic algorithm with a two-dimensional discrete Haar wavelet transform. The quality of target coverage is measured based on the Boolean matrix sum. The initial random population matrix applied crossover and mutation, and later, 2D Haar wavelet transform was applied for local search.

The authors of [6] added modified GA with 2D Lifting wavelet transform for fast adjustment of GA solutions. Initially, row-wise computation of Lazy lifting transform and then column-wise Lazy lifting operation adjusts the GA random population matrix of sensor positions [12]. Instead of Lazy lifting, the authors of [14] presented the 2D-CDF 5/3 lifting wavelet transform for fast computation. The authors of [18] formulated the target coverage for the mathematical model. The coverage and connectivity depended on the energy spend on individual sensors. The Hop Distance and Send Request algorithms are used to optimize the energy between the source that covers the target to sink node. The Find Covers algorithm generates the cover sets of sensors covered targets.

In this chapter, the proposed improved genetic algorithm (GA) produces the initial random sensor position in the form of a population matrix. This random population is applied to single-point crossover and mutation operation. The resultant matrix is applied to local enhancement operation using a two-dimensional discrete Haar lifting wavelet transform, which improves the child matrices of sensor positions. First, the resultant matrices are applied for fitness function, which find the coverage and connectivity of each sensor node. Second, the quality of target coverage is measured based on the Boolean matrix. The design of the proposed improved GA with 2D DHLWT is expressed in Sections 2.3 and 2.4.

2.3   PROBLEM FORMULATION

Let $S = \{S_1, S_2, \ldots, S_i, \ldots S_n\}$ be the set of n available sensor nodes in the area to supervise set of m available targets, whereas $T = \{T_1, T_2, \ldots, T_k, \ldots T_m\}$. The set of n available sensor nodes is less than the set of m available targets. When $n \geq m$ maybe the sensor nodes are placed very close to each target and redundant target cover occurs mostly. Each sensor S_i has sensing range r_s and communication range r_c placed (x_i, y_i) in the two-dimensional area $A \times A$, to examine the target T_k located in (x_k, y_k) the similar region. If the sensor node S_i covers the target T_k, that satisfies Equation (2.1). Suppose sensor node S_j is placed (x_j, y_j) in the same two-dimensional region $A \times A$, whereas the sensor node S_i establishes a connection to the sensor node S_j to satisfy Equation (2.2).

$$r_s^2 \geq \left(x_i - x_k\right)^2 + \left(y_i - y_k\right)^2, 1 \leq i \leq n, 1 \leq k \leq m \tag{2.1}$$

$$r_c^2 \geq \left(x_i - x_j\right)^2 + \left(y_i - y_j\right)^2, 1 \leq \left(i,j\right) \leq n, i \neq j \tag{2.2}$$

Every sensor node coverage and connectivity can be discovered by using Equations (2.1) and (2.2), and it can be exemplified in the form of a Boolean matrix in Equations (2.3) and (2.4).

$$\Phi_{n \times m} = \begin{matrix} S_1 \\ S_2 \\ \vdots \\ \vdots \\ S_i \\ S_n \end{matrix} \begin{bmatrix} \Phi_{11} & \Phi_{12} & \cdots & \Phi_{1m} \\ \Phi_{21} & \Phi_{22} & \cdots & \Phi_{2m} \\ \cdots & \cdots & \cdots & \cdots \\ \cdots & \cdots & \cdots & \cdots \\ \Phi_{i1} & \Phi_{i2} & \cdots & \Phi_{ik} \\ \Phi_{n1} & \Phi_{n2} & \cdots & \Phi_{nm} \end{bmatrix} \tag{2.3}$$

$$Where\ \Phi_{ik} = \begin{cases} 1, if\ S_i\ monitor\ T_k\ and\ S_i\ connect\ S_j \\ 0, otherwise. \end{cases} \tag{2.4}$$

This matrix represents the Boolean values of the sensors; each sensor either covers the target or connects to another sensor node. Each row represents the sensor node S_i whether it covers targets or not. The coverage and connect matrix are to be calculated as column sum matrix of each target cover sets.

$$\tau = \left[\tau_1\ \tau_2\ \tau_3 \ldots \tau_k \ldots \tau_m\right] where\ \tau_k = \sum_{i=1}^{i=n} \Phi_{ik} \tag{2.5}$$

The quality of the total target point covers is classified as the total number of targets covered by the total number of sensors deployed in the environmental region for monitoring purposes.

$$QoC\left(\%\right) = \frac{\left(\sum_{k=1}^{m}\left(\dfrac{\tau_k}{n}\right)\right)}{m} \times 100 \tag{2.6}$$

The sample network region 100×100 consist of a total number of available sensors (n = 8) and available number of targets (m = 16), where Figure 2.1 represents sample random deployment of target points and Figures 2.2 and 2.3 represent the sample random deployment of sensor nodes in the population matrix. The coverage of every

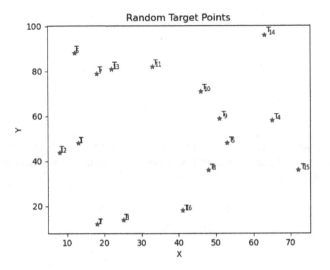

FIGURE 2.1 Randomly deployed target points.

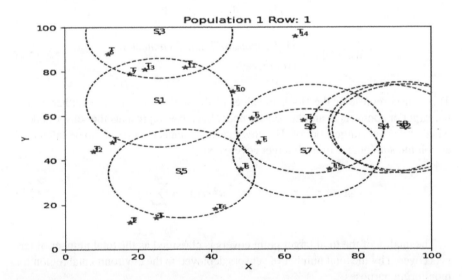

FIGURE 2.2 Randomly deployed sensor nodes in Pop1.

population row is measure by using Equations (2.1) and (2.2) where it covers the target and connected with sensor nodes. The population 1, row 1 coverage sets are $S_1 = \{T_7, T_{11}, T_{13}\}$, $S_2 = \{\}$, $S_3 = \{T_6, T_7, T_{11}, T_{13}\}$, $S_4 = \{\}$, $S_5 = \{T_8, T_{16}\}$, $S_6 = \{T_4, T_5, T_9, T_{15}\}$, $S_7 = \{T_4, T_5, T_8, T_{15}\}$, and $S_8 = \{\}$. The population 2 row 1 coverage sets are $S_1 = \{T_5, T_9, T_{10}\}$, $S_2 = \{\}$, $S_3 = \{T_1, T_{12}\}$, $S_4 = \{T_{10}, T_{11}, T_{13}\}$, $S_5 = \{T_1, T_{12}\}$, $S_6 = \{T_7, T_{10}, T_{11}, T_{13}\}$, $S_7 = \{T_1, T_{12}\}$, and $S_8 = \{T_5, T_8, T_9\}$. Population 1, row 1 and population 2, row 1 have the quality of total target coverage as 62.5% and 56.25% respectively.

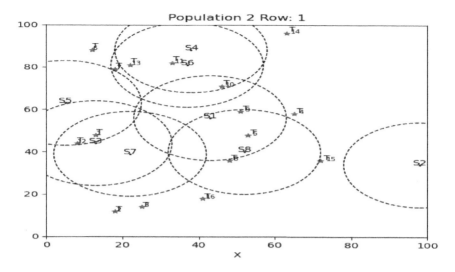

FIGURE 2.3   Randomly deployed sensor nodes in Pop2.

2.4   HAAR LIFTING SCHEME

Over the past few years, many studies have been conducted on one-dimensional and two-dimensional discrete wavelet transform for picture and video signals [19,20]. The time complexity of the arithmetic operation has sped up when using the lifting scheme-based wavelet transform. Analysis and synthesis filter into polyphase components while applying decomposition. The following coefficients are Haar filter banks analysis and synthesis values.

$$h_a(z) = (1 + z^{-1})/2 \, analysis \, Low \, Pass \, Filter \tag{2.7}$$

$$h_s(z) = (1 + z^{-1}) \, synthesis \, Low \, Pass \, Filter \tag{2.8}$$

$$g_a(z) = (z^{-1} - 1) \, analysis \, High \, Pass \, Filter \tag{2.9}$$

$$g_s(z) = (z^{-1} - 1)/2 \, synthesis \, High \, Pass \, Filter \tag{2.10}$$

The polyphase matrix can be computed while executing the Euclidean algorithm [21,22]. The computation of the one-dimensional output coefficients is given in Equation (2.11) using a polyphase matrix.

$$P(z) = \begin{bmatrix} h_e(z) & g_e(z) \\ h_o(z) & g_o(z) \end{bmatrix} s = \begin{bmatrix} 1 & -1/2 \\ 1 & 1/2 \end{bmatrix} = \begin{bmatrix} 1 & 0 \\ 1 & 1 \end{bmatrix} \begin{bmatrix} 1 & -1/2 \\ 0 & 1 \end{bmatrix} \tag{2.11}$$

$h_e(z)$ is the even coefficient value and $h_o(z)$ is the odd coefficient values of analysis and synthesis filter banks respectively.

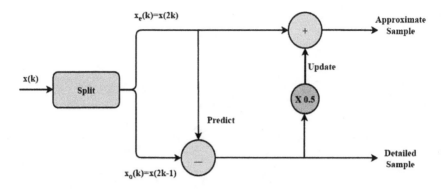

FIGURE 2.4   Haar lifting wavelet transform.

$$Split : \alpha_0\left(k\right) = x_e\left(k\right) \text{ Even Component}, \delta_0\left(k\right) = x_o\left(k\right) \text{ Odd Component} \quad (2.12)$$

$$Predict:\ \delta\left(k\right) = \delta_0\left(k\right) - \alpha_0\left(k\right) \quad (2.13)$$

$$Update:\ \alpha\left(k\right) = \alpha_0\left(k\right) + \frac{\delta\left(k\right)}{2} \quad (2.14)$$

The input signal x(k) splits into even x_e(k) = x(2k) and odd x_o(k) = x(2k − 1), the detailed components are derived from the even samples called prediction, and the approximate component is derived from the detailed components called the updating operation [23]. Figure 2.4 shows the Haar lifting wavelet transform.

Normally, 2-D wavelet transforms utilized while applying the 1-D transform in the row-wise and then in column-wise or vice versa represents in Figure 2.5. The given

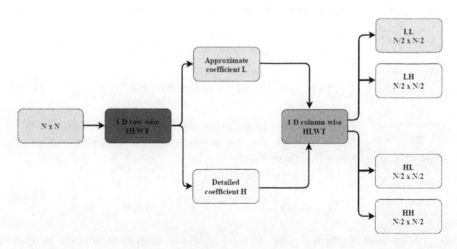

FIGURE 2.5   2-D Haar lifting computation.

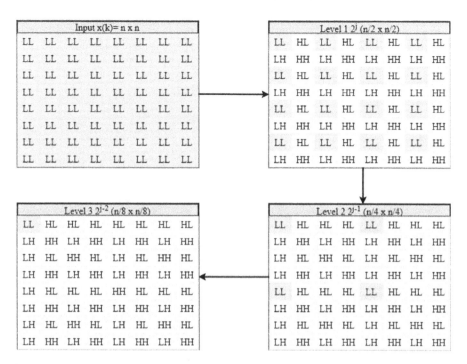

FIGURE 2.6   2-D Haar lifting decomposition.

$n \times n$ matrix has $2^j = n$ components and it can be computed row-wise and then column-wise level-1 decomposition. After level-1 computation reduces the matrix into $n/2 \times n/2$ having $2^{j-1} = n/2$ components. The same thing will be continued in the next level of decomposition using approximate signal until the approximate data in the matrix is reached. The process is represented in Figure 2.6.

2.5   GA-2D HAAR LIFTING OPTIMAL SENSOR PLACEMENT

The solution to be optimized uses the more generalized genetic algorithm which combines a few of the evolutionary algorithms. The main kind of GA is used to change the initial solution into an optimized solution utilizing the crossover, mutation, and some local enhancements operation. Initially, GA generates the random population matrices, which contain the sensor coordinates. Then, it exchanges the genes between two chromosomes (different populations) followed by updating the random gene values (mutation). The quality of the sensor position is applied to the local improvement of two-dimensional Haar lifting wavelet transform, which adjusts the sensor coordinates into an optimum position to cover the maximum number of targets and connect to another sensor node. The super subsistence operator selects the best-optimized solution from the two different populations to child matrices for the next iteration. This process is repeated a maximum number of times to get an optimized position of sensor nodes. The main framework of this proposed algorithm for sensor deployment is given in Algorithm 2.1.

ALGORITHM 2.1   GENETIC ALGORITHM WITH 2D HLWT

Input: Random target position (x_k, y_k), n, m, r_s, r_c.
Output: Optimized sensor node coordinate position.
Start Algorithm
Limit=int (input ("Enter total number of Population
Pop [] = Random group of initial Population
for i=1: limit
       Crossover_pop = *func_Crossover(Pop[i-1],Pop[i])*
       Mutation_pop   = *func_Mutation(Pop[i])*
       Child_pop   =*localEnhance2DHLWT(Pop[i])*
       Pop[i]   =*func_Survival(Child_pop,Pop[i])*
End algorithm
Deploy the sensor coordinate positions

---

1. *Representation*: every sensor position can be signified in the form of population matrices called chromosomes. The size of each chromosome or population is a n × n matrix, where n is the total sum of sensor nodes. Every chromosome value is generated randomly in range 0 to A × A, where A is the size of the environmental area. Every row in the population matrix represents the sensor place.

$$\rho = \begin{bmatrix} \rho_{11} & \rho_{12} & \cdots & \cdots & \rho_{1n} \\ \rho_{21} & \rho_{22} & \cdots & \cdots & \rho_{2n} \\ \cdots & \cdots & \cdots & \cdots & \cdots \\ \cdots & \cdots & \cdots & \cdots & \cdots \\ \rho_{n1} & \rho_{n2} & \cdots & \cdots & \rho_{nn} \end{bmatrix} \qquad (2.15)$$

2. *Initial population*: a group of initial population is produced randomly to search optimum sensor position. However, ρ_1 and ρ_2 are two different population samples generated randomly. Here, every population matrix contains n number of rows and n number of columns, which is the number of available sensor nodes (n = 8). The chromosome of the population matrix assigns the random values between 0 and 100 × 100.

$$\rho_1 = \begin{bmatrix} 666 & 9354 & 2697 & 8754 & 3234 & 6754 & 6643 & 9255 \\ 1601 & 9321 & 3452 & 7865 & 1203 & 5472 & 6467 & 9998 \\ 2314 & 8764 & 4224 & 9876 & 4215 & 9811 & 1567 & 8934 \\ 1374 & 9963 & 1945 & 7185 & 1045 & 7892 & 708 & 7853 \\ 2410 & 7894 & 1345 & 7646 & 1035 & 8532 & 2345 & 8679 \\ 657 & 3453 & 1586 & 8534 & 3624 & 7447 & 2323 & 7321 \\ 578 & 5645 & 3245 & 7821 & 4221 & 8087 & 5867 & 8682 \\ 4532 & 9898 & 2345 & 8975 & 2357 & 8769 & 973 & 5470 \end{bmatrix}$$

$$
P_2 = \begin{bmatrix}
4356 & 9834 & 1344 & 3888 & 563 & 3781 & 2239 & 5240 \\
850 & 4680 & 3057 & 9034 & 3456 & 7896 & 457 & 1067 \\
5003 & 8766 & 987 & 9793 & 4052 & 6745 & 3221 & 5305 \\
2323 & 6778 & 1989 & 9072 & 2234 & 7878 & 1043 & 4590 \\
5643 & 7908 & 3498 & 5460 & 3408 & 6787 & 5467 & 8754 \\
4367 & 8777 & 3421 & 7049 & 1098 & 9833 & 1005 & 4789 \\
1877 & 5329 & 4280 & 8790 & 3419 & 5438 & 8776 & 9983 \\
3453 & 7853 & 3287 & 9342 & 5980 & 8092 & 2453 & 4490
\end{bmatrix}
$$

3. *Fitness function*: to search optimal space for the way out, the fitness function is utilized to find the fitness of each chromosome, whereas each chromosome is converted into a sensor coordinate place by using Equations (2.3) through (2.6).

$$
\begin{pmatrix} x_i \\ y_i \end{pmatrix} = \begin{pmatrix} abs\left(\dfrac{P_{ij}}{A}\right) \\ \left(P_{ij} \bmod (A)\right) \end{pmatrix} \tag{2.16}
$$

The entry of chromosome rows has represented the available sensor coordinate position. However, the first row P_1 and P_2 is converted into a sensor coordinate position using Equation (2.16).

$$
P_1 = \begin{pmatrix} x \\ y \end{pmatrix} = \begin{pmatrix} 6 & 93 & 26 & 87 & 32 & 67 & 66 & 92 \\ 66 & 54 & 97 & 54 & 34 & 54 & 43 & 55 \end{pmatrix}
$$

$$
P_2 = \begin{pmatrix} x \\ y \end{pmatrix} = \begin{pmatrix} 43 & 98 & 13 & 38 & 5 & 37 & 22 & 52 \\ 56 & 34 & 44 & 88 & 63 & 81 & 39 & 40 \end{pmatrix}
$$

This set of sensors are virtually deployed in the region to supervise the given set of targets. The quality of each population matrices is evaluated using Equation (2.6).

4. Crossover: To exchange genes among the two different chromosomes is recognized as crossover. GA randomly obtain a value from1 to 8 (n). Let us take over a randomly generated the single-point crossover is 4. Consequently, the offspring (os_1) is formed from the first 4 columns from P_1 and the last 4 columns from P_2. Similarly, offspring (os_2) is the first 4 columns from P_2 and last 4 columns from P_1.

$$
os_1 = \begin{bmatrix}
666 & 9354 & 2697 & 8754 & 563 & 3781 & 2239 & 5240 \\
1601 & 9321 & 3452 & 7865 & 3456 & 7896 & 457 & 1067 \\
2314 & 8764 & 4224 & 9876 & 4052 & 6745 & 3221 & 5305 \\
1374 & 9963 & 1945 & 7185 & 2234 & 7878 & 1043 & 4590 \\
2410 & 7894 & 1345 & 7646 & 3408 & 6787 & 5467 & 8754 \\
657 & 3453 & 1586 & 8534 & 1098 & 9833 & 1005 & 4789 \\
578 & 5645 & 3245 & 7821 & 3419 & 5438 & 8776 & 9983 \\
4532 & 9898 & 2345 & 8975 & 5980 & 8092 & 2453 & 4490
\end{bmatrix}
$$

$$os_2 = \begin{bmatrix} 4356 & 9834 & 1344 & 3888 & 3234 & 6754 & 6643 & 9255 \\ 850 & 4680 & 3057 & 9034 & 1203 & 5472 & 6467 & 9998 \\ 5003 & 8766 & 987 & 9793 & 4215 & 9811 & 1567 & 8934 \\ 2323 & 6778 & 1989 & 9072 & 1045 & 7892 & 708 & 7853 \\ 5643 & 7908 & 3498 & 5460 & 1035 & 8532 & 2345 & 8679 \\ 4367 & 8777 & 3421 & 7049 & 3624 & 7447 & 2323 & 7321 \\ 1877 & 5329 & 4280 & 8790 & 4221 & 8087 & 5867 & 8682 \\ 3453 & 7853 & 3287 & 9342 & 2357 & 8769 & 973 & 5470 \end{bmatrix}$$

5. *Mutation*: moreover, randomly update some chromosome gene values for the ideal solution. GA produces a random mutation point as 4 and 6 for os_1 and os_2, respectively. Therefore, the 4th column of os_1 and 6th column of os_2 are redeveloped again.

$$os_1 = \begin{bmatrix} 666 & 9354 & 2697 & 8975 & 563 & 3781 & 2239 & 5240 \\ 1601 & 9321 & 3452 & 8743 & 3456 & 7896 & 457 & 1067 \\ 2314 & 8764 & 4224 & 7865 & 4052 & 6745 & 3221 & 5305 \\ 1374 & 9963 & 1945 & 6745 & 2234 & 7878 & 1043 & 4590 \\ 2410 & 7894 & 1345 & 5643 & 3408 & 6787 & 5467 & 8754 \\ 657 & 3453 & 1586 & 9898 & 1098 & 9833 & 1005 & 4789 \\ 578 & 5645 & 3245 & 7668 & 3419 & 5438 & 8776 & 9983 \\ 4532 & 9898 & 2345 & 6865 & 5980 & 8092 & 2453 & 4490 \end{bmatrix}$$

$$os_2 = \begin{bmatrix} 4356 & 9834 & 1344 & 3888 & 3234 & 8976 & 6643 & 9255 \\ 850 & 4680 & 3057 & 9034 & 1203 & 5756 & 6467 & 9998 \\ 5003 & 8766 & 987 & 9793 & 4215 & 7897 & 1567 & 8934 \\ 2323 & 6778 & 1989 & 9072 & 1045 & 9865 & 708 & 7853 \\ 5643 & 7908 & 3498 & 5460 & 1035 & 8797 & 2345 & 8679 \\ 4367 & 8777 & 3421 & 7049 & 3624 & 6789 & 2323 & 7321 \\ 1877 & 5329 & 4280 & 8790 & 4221 & 9898 & 5867 & 8682 \\ 3453 & 7853 & 3287 & 9342 & 2357 & 9867 & 973 & 5470 \end{bmatrix}$$

6. Local Enhancement: 2D Haar lifting wavelet transform is applied in the os_1 and os_2 for multi-level decomposition to produce covers to the maximum number of targets and connect to other sensor nodes using deterministic sensor placement. After applying 2DHLWT, the os_1 and os_2 are to be a $child_1$ and $child_2$, correspondingly. Further, some points are negative and out of the region, so as to apply some threshold formula to bring all the sensors into the area.

$$\rho_{ij} = \left(\left(abs \left(\rho_{ij} \right) \right) \bmod \left(A \times A \right) \right) \qquad (2.17)$$

$$child_1 = \begin{bmatrix} 4863 & 8204 & 161 & 5785 & 379 & 3829 & 1680 & 1806 \\ 451 & 968 & 262 & 987 & 3504 & 1222 & 2978 & 2391 \\ 202 & 7520 & 1140 & 4221 & 1296 & 4169 & 14 & 2816 \\ 130 & 2139 & 1700 & 1159 & 343 & 2951 & 1447 & 1463 \\ 490 & 4140 & 441 & 6305 & 2772 & 6057 & 208 & 3536 \\ 3097 & 2688 & 2248 & 4014 & 368 & 5356 & 4214 & 497 \\ 986 & 5217 & 1147 & 4472 & 936 & 2066 & 971 & 1622 \\ 4104 & 299 & 852 & 97 & 2608 & 93 & 5908 & 830 \end{bmatrix}$$

$$child_2 = \begin{bmatrix} 5531 & 4654 & 428 & 4261 & 306 & 5148 & 1154 & 3072 \\ 4330 & 1648 & 3430 & 3433 & 2626 & 1189 & 284 & 919 \\ 959 & 4109 & 342 & 7945 & 1181 & 6251 & 4288 & 7256 \\ 2334 & 692 & 141 & 1723 & 601 & 5138 & 970 & 222 \\ 100 & 3338 & 10 & 2795 & 872 & 5464 & 616 & 5666 \\ 204 & 2145 & 756 & 1666 & 291 & 4597 & 690 & 1336 \\ 239 & 3926 & 3614 & 5283 & 803 & 6594 & 1444 & 3656 \\ 2050 & 948 & 221 & 1545 & 948 & 1833 & 4053 & 1682 \end{bmatrix}$$

Finally, $child_1$ and $child_2$ matrices rows are converted into sensor place as defined in the fitness function and represented in Figures 2.7 and 2.8. Thus, sensors can be placed in this optimum location to cover the maximum number of targets and be able to connect with other sensor nodes for better network connectivity. The $child_1$, row 1

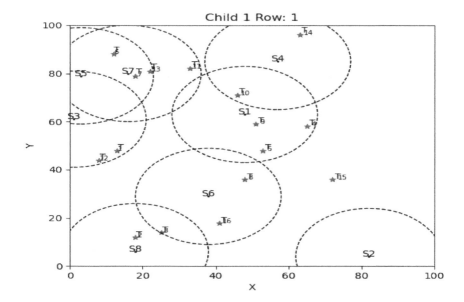

FIGURE 2.7   Child 1 row 1 sensor deployment.

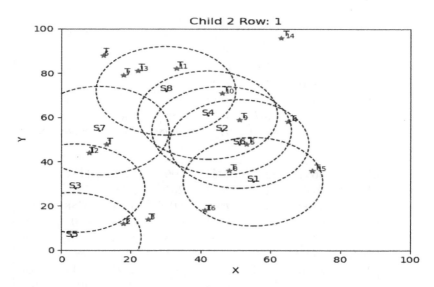

FIGURE 2.8 Child 2 row 1 sensor deployment.

coverage sets are $S_1 = \{T_4, T_5, T_9, T_{10}\}$, $S_2 = \{\}$, $S_3 = \{T_1, T_{12}\}$, $S_4 = \{T_{10}, T_{14}\}$, $S_5 = \{T_6, T_7, T_{13}\}$, $S_6 = \{T_3, T_8, T_{16}\}$, $S_7 = \{T_6, T_7, T_{11}, T_{13}\}$, and $S_8 = \{T_2, T_3\}$. The child$_2$, row 1 coverage sets are $S_1 = \{T_5, T_8, T_{15}, T_{16}\}$, $S_2 = \{T_4, T_5, T_8, T_9, T_{10}\}$, $S_3 = \{T_{12}\}$, $S_4 = \{T_5, T_9, T_{10}\}$, $S_5 = \{T_2\}$, $S_6 = \{T_4, T_5, T_8, T_9\}$, $S_7 = \{T_1, T_{12}\}$, and $S_8 = \{T_7, T_{10}, T_{11}, T_{13}\}$. The quality of total target coverage of child$_1$, row 1 and child$_2$, row 1 is 93.75% and 81.25%.

2.6 SIMULATION RESULTS

The serious simulation work was conducted in a MATLAB® environment using Windows 7, i5 processor, 2.76 GHz, 4 GB ram for evaluating the proposed method. The simulation work was carried out using randomly generated sensors count ranging from 20 to 150; total target counts ranged from 40 to 500. Each sensor has the same sensing ranges, varied from 50 to 150 and the communication range varied from 100 to 300. The initial population 100, simulation region size 100×100, crossover point 1, and mutation point 1 are measured in this simulation. The simulation work consists of different sensing range, and scenarios are formulated as in the given format, $m = \sqrt{2}n$, $m = \sqrt{4}n$, where n and m are sensors counts and target counts respectively.

Tables 2.1 and 2.2 shows that Qoc generated for sensing of the sensor is 50 and with various sensor target ratios as $m = \sqrt{2}n$ and $m = \sqrt{4}n$ respectively. The sensors and targets are deployed in 100 x 100 region. The available sensor counts vary from 32 to 128, whereas it is compared with random, GA, and proposed methods. The average Qoc in the proposed method of Table 2.1 is 88.89%, whereas random deployment average Qoc is 72.26% and GA has 80.37%. Similarly, the Qoc of Table 2.2 of proposed GA + 2DDHLWT is always greater than 90%.

TABLE 2.1

Qoc of Sensing Range 50 and m = $\sqrt{2}n$

Sensors: n	32		64		128	
Algorithms	Qoc	Qoc (%)	Qoc	Qoc (%)	Qoc	Qoc (%)
Random	28	62.222	68	75.556	143	79.005
GA	30	66.667	76	84.444	163	90.005
GA+2DDHLWT	36	80.000	82	91.111	173	95.580

TABLE 2.2

Qoc of Sensing Range 50 and m = $\sqrt{4}n$

Sensors: n	32		64		128	
Algorithms	Qoc	Qoc (%)	Qoc	Qoc (%)	Qoc	Qoc (%)
Random	48	75.000	102	79.687	216	84.375
GA	53	82.812	116	90.625	245	95.703
GA+2DDHLWT	59	92.187	120	93.750	248	96.875

Figures 2.9 and 2.10 show that the number of sensors versus the number of target points with sensing range 75 with different target ratios. The result is obtained in the 100 × 100 region, whereas the sensing range of every sensor is a maximum value. Compared to random and GA, the proposed method has a maximum number of Qoc.

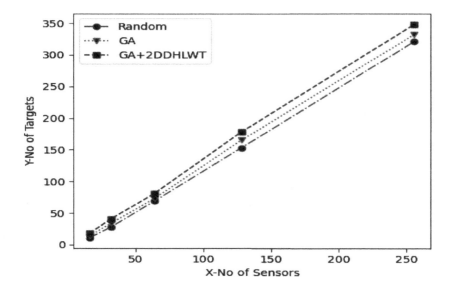

FIGURE 2.9   Qoc created for sensing range 75 and m = $\sqrt{2}n$.

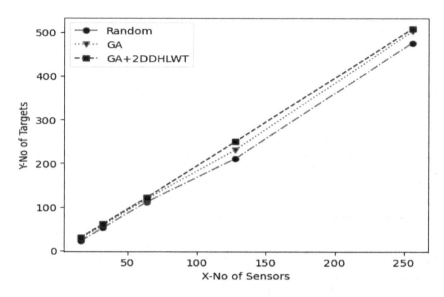

FIGURE 2.10 Qoc created for sensing range 75 and m = $\sqrt{4}$n.

Figures 2.11 and 2.12 show that Qoc (%) versus the number of sensors and sensors has sensing of 150 and targets are m = $\sqrt{2}$n and m = $\sqrt{4}$n. Figures 2.11 and 2.12 show that the proposed method has higher Qoc (%) compared to random and GA deployment. It implies that the solution produced by GA + 2DDHLWT is better than the random placement. It is expected to be the dilation and the translation of the discrete Haar wavelet, which totally spread the sensors in the entire region.

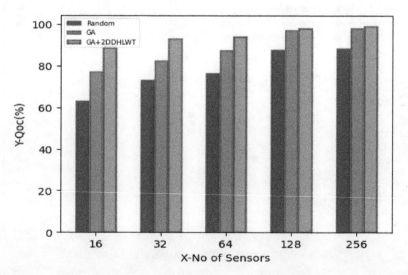

FIGURE 2.11 Qoc (%) created for sensing range 150 and m = $\sqrt{2}$n.

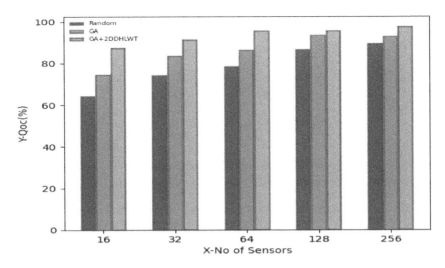

FIGURE 2.12   Qoc (%) created for sensing range 150 and m = $\sqrt{4}$n.

2.7   CONCLUSION

The procedure of WSN solutions for real-time functions, as it obtains the available resources and their system structure setup, is based on the deployment of nodes and target points. The main objectives of the target points coverage process depend on searching of optimal sensor cover sets which can extend the network lifetime. This chapter has mentioned two main subdivisions to discover the optimum location of sensor nodes, which covers the maximum number of target points. The new techniques called improved GA to identify the estimated optimal solution and 2-DDHLWT efficiently formulate GA into an optimal solution while applying multilevel decomposition. Moreover, the sensor nodes move to the optimal location to cover maximum target points and established connections with other sensors. A set of simulation works carried out in the proposed methodology and results have been validated with different scenarios. Yet, the proposed work is analyzed against the homogeneous WSNs in which every sensor has similar sensing capability. The simulation outcomes of the proposed technique have confirmed the maximum target points covered and the connectivity of nodes that are compared with available existing algorithms. Additional simulation setup is required in the future to measure the network lifetime, as is a different lifting wavelet such as a genetic algorithm with a lifting scheme.

REFERENCES

1. Chia-Pang Chen, Subhas Chandra Mukhopadhyay, Cheng-Long Chuang, Tzu-Shiang Lin, Min-Sheng Liao, Yung-Chung Wang, and Joe-Air Jiang, "A hybrid memetic framework for coverage optimization in wireless sensor networks," *IEEE Transactions on Cybernetics*, Vol. 45, No. 10, pp. 2309–2321, 2015.
2. Ying Liu, Kwan-Wu Chin, Changlin Yang, and Tengjiao He, "Nodes deployment for coverage in rechargeable wireless sensor networks," *IEEE Transactions on Vehicular Technology*, Vol. 68, No. 6, pp. 6064–6073, 2019.

3. Hicham Ouchitachen, Abdellatif Hair, and Najlae Idrissi, "Improved multi-objective weighted clustering algorithm in Wireless Sensor Network," *Egyptian Informatics Journal*, Vol. 18, No. 1, pp. 1–10, 2016.

4. Julio Vilela, Zendai Kashino, Raymond Ly, Goldie Nejat, and Beno Benhabib, "A dynamic approach to sensor network deployment for mobile-target detection in unstructured, expanding search areas," *IEEE Sensors Journal*, Vol. 16, No. 11, pp. 4405–4417, 2016.

5. Mohamed Elhoseny, Alaa Tharwat, Ahmed Farouk, and Aboul Ella Hassanien, "k-coverage model based on genetic algorithm to extend WSN lifetime," *IEEE Sensors Letter*, Vol. 1, No. 4, 2017.

6. T. Ganesan and Pothuraju Rajarajeswari, "Genetic algorithm approach improved by 2D lifting scheme for sensor node placement in optimal position," *Second International Conference on Intelligent Sustainable Systems*, Vol. 1, pp. 104–109, 2019.

7. P. Vijayaraju, B. Sripathy, D. Arivudainambi, and S. Balaji, "Hybrid memetic algorithm with two-dimensional discrete Haar wavelet transform for optimal sensor placement," *IEEE Sensors Journal*, Vol. 17, No. 7, pp. 2267–2278, 2017.

8. Kasilingam Rajeswari and Subbu Neduncheliyan, "Genetic algorithm-based fault-tolerant clustering in wireless sensor network," *IET Communication*, Vol. 11, No. 12, pp. 1927–1932, 2017.

9. Mustapha Reda Senouci, and Abdelhamid Mellouk, "A robust uncertainty-aware cluster-based deployment approach for WSNs: Coverage, connectivity, and lifespan," *Journal of Network and Computer Applications*, Vol. 146, No. 2, pp. 1–12, 2019.

10. Osama Moh'd Alia and Alaa Al-Ajouri, "Maximizing wireless sensor network coverage with minimum cost using harmony search algorithm," *IEEE Sensors Journal*, Vol. 17, No. 3, pp. 882–896, 2017.

11. Chien-Chih Liao and Chuan-Kang Ting, "A novel integer-coded memetic algorithm for the set k-cover problem in wireless sensor networks," *IEEE Transaction Cybernetics*, Vol. 48, No. 8, pp. 2245–2257, 2018.

12. T. Ganesan and Pothuraju Rajarajeswari, "Genetic algorithm based optimization to improve the cluster lifetime by optimal sensor placement in WSN's," *International Journal of Innovative Technology and Exploring Engineering*, Vol. 8, No. 8, pp. 3400–3408, 2019.

13. Nguyen Thi Hanh, Huynh Thi Thanh Binh, Nguyen Xuan Hoai, and Marimuthu Swami Palamiswami, "An efficient genetic algorithm for maximizing area coverage in wireless sensor networks," *Information Sciences*, Vol. 488, No. 1, pp. 58–75, 2019.

14. T. Ganesan, Pothuraju Rajarajeswari, Soumya Ranjan Nayak, and Amandeep Singh Bhatia, "A novel genetic algorithm with CDF5/3 filter-based lifting scheme for optimal sensor placement," *International Journal of Innovative Computing and Applications*, Vol. 12, No. 4, pp. 1–10, 2021.

15. Phi Le Nguyen, Nguyen Thi Hanh, Nguyen Tien Khuong, Huynh Thi Thanh Binh, and Yusheng Ji, "Node placement for connected target coverage in wireless sensor networks with dynamic sinks," *Pervasive and Mobile Computing*, Vol. 59, No. 2, pp. 1–21, 2019.

16. Weiqiang Wang, "Deployment and optimization of wireless network node deployment and optimization in smart cities," *Computer Communications*, Vol. 22, No. 1, pp. 1–11, 2020.

17. Wafa Barkhoda and Hemmat Sheikhi, "Immigrant imperialist competitive algorithm to solve the multi-constraint node placement problem in target-based wireless sensor networks," *Ad Hoc Networks*, Vol. 16, No. 2, pp. 1–21, 2020.

18. Anvesha Katti, "Target coverage in random wireless sensor networks using cover sets," *Journal of King Saud University—Computer and Information Sciences*, Vol. 6, No. 1, pp. 1–13, 2019.

19. Yeong-Kang Lai, Lien-Fei Chen, and Yui-Chih Shih, "A high-performance and memory-efficient VLSI architecture with parallel scanning method for 2-D lifting-based discrete wavelet transform," *IEEE Transactions on Consumer Electronics*, Vol. 55, No. 2, pp. 400–407, 2009.

20. Xin Tian, Lin Wu, Yi-Hua Tan, and Jin-Wen Tian, "Efficient multi-input/multi-output VLSI architecture for two-dimensional lifting-based discrete wavelet transform," *IEEE Transactions on Computers*, Vol. 60, No. 8, pp. 1207–1211, 2011.

21. Eric J. Balster, Benjamin T. Fortner, and William F. Turri, "Integer computation of lossy JPEG2000 compression," *IEEE Transactions on Image Processing*, Vol. 20, No. 8, pp. 2386–2391, 2011.

22. Md. Mehedi Hasan, and Khan A. Wahid, "Low-cost architecture of modified Daubechies lifting wavelets using integer polynomial mapping," *IEEE Transactions on Circuits and Systems II*, Vol. 2, No. 10, pp. 1–5, 2016.

23. Serwan Ali Bamerni and Ahmed Kh. Al-Sulaifanie, "An efficient non-separable architecture for Haar wavelet transform with lifting structure," *Journal of Microprocessors and Microsystems*, Vol. 71, No. 1, pp. 1–7, 2019.

3 Lifetime Enhancement of Wireless Sensor Network Using Artificial Intelligence Techniques

Jayashree Dev and Jibitesh Mishra

College of Engineering and Technology, Bhubaneswar, Odisha, India

CONTENTS

3.1 INTRODUCTION

A wireless sensor network (WSN) is a collection of interconnected, interrelated, and distributed sensor nodes that are designed to sense and measure information from the environment and send them to base station for processing. Besides the sensor nodes, it contains different base stations. Base stations are meant for processing of collected data. Result of data processing is used in different applications. Today, we can see a lot of applications of WSN, including battlefield surveillance, health application, research and education, and agriculture (Figure 3.1).

The sensor nodes are very tiny in shape and are battery powered. Batteries may be rechargeable or non-rechargeable. Rechargeable batteries are used in the WSN, which is deployed in an area where human intervention is possible after deployment

DOI: 10.1201/9781003145028-3

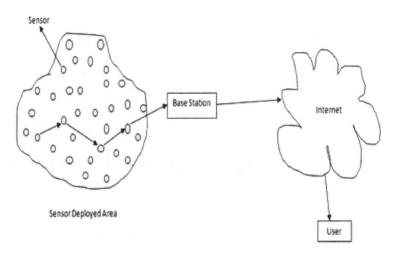

FIGURE 3.1 Wireless sensor network.

of the network. For example, in a smart home system, chargeable batteries can be used. This type of network has very few real-time applications. In the areas which are not easily accessible to the human being, where it is quite impossible to deploy a human being, WSNs with non-rechargeable batteries are used. For example, in case of an earthquake prediction system that is designed to monitor the probability of occurrence of earthquake on a sea bed, a non-rechargeable sensor network is deployed. The objective of this system is to protect property and lives of people by issuing alerts to take precaution. Similarly, in a lot of application areas, there is a demand of non-rechargeable battery-based WSN. This demand is growing with the growing application of the WSN in different fields [1].

The non-rechargeable battery-based WSN has a lot of constraints including the limited battery power as the main constraint. A huge volume of continuous data are sensed and generated by the sensors which are to be reached at base station for processing through multi-hop communication [2].

Battery power consumption occurs when the sensor node is busy with sensing and collecting information, in-node processing, sending the information to the neighboring node, receiving information from the neighboring node, processing overhead information including control overhead information. Even in case of a dynamic WSN, each time, the network configuration changes with the change in the place of sensors and base stations. In this case, for coordination between sensor nodes and base stations, there is a need of exchange of location information among them which increases the amount of communications in the network and hence results in more overhead information [3].

When the volume of information is higher and the amount of end-to-end communication is greater, there is a great chance of early battery depletion which causes network death. In case of non-rechargeable battery-based WSN, it is very crucial to save the network from pre-timely death because with the death of the network, our good purpose dies as well. Even when most of the sensor nodes of a WSN dies due to loss of battery power, the network behaves peculiarly, resulting in incorrect output. The raw data sensed by the sensors undergoes in-node processing and then is sent to the neighboring node in the aim of being received at the base station. If all the neighboring nodes of a sensor node present within its communication range are dead, then the collected information can't be sent to the base station/sink. This is called the partitioning of the network. In case of network partitioning, though the network is not completely dead, it surely affects network performance [4].

To avoid early network death, care should be taken to minimize power consumption at the time of occurrence of different operations at the sensor node like sensing, in-node processing, and sending/receiving messages from neighboring node/base station. When the traditional way of data processing is used by sensor nodes and the base station, it takes more time for processing and generates a lot of overhead. In order to improve the performance of the network, traditional data processing techniques can be replaced with the artificial intelligence technique which has the ability of doing the task using humanlike intelligence. Artificial intelligence techniques can process a large volume of data in very little time and provides a more accurate result on which one can rely and proceed with the next level task [5,6].

3.2 ISSUES IN WIRELESS SENSOR NETWORK

Different issues that exist in a WSN are [7] (Figure 3.2):-

- *Network topology*: in WSN, sensor nodes are deployed in a distributed fashion over a target region. The network does not fit into exactly any topology. In order to facilitate smooth communication among sensor nodes and to avoid long-distance communication, the sensor nodes collectively form a suitable

FIGURE 3.2 Different issues in wireless sensor network.

structure which is useful in getting high throughput as well as high accuracy in output.

- *Limited resource*: sensor nodes are tiny components consisting of a sensing unit, processing unit, power unit and transceiver unit. The function of the sensing unit is to collect and measure data from the environment. The processing unit is meant for in-node processing of collected data, removing redundant data from it, and compressing those data. The power unit is responsible for managing power supply in the node and facilitates all operations that are done in a node. The transceiver unit is responsible for facilitating node-to-node communication. The sensor nodes are battery powered and the battery may be chargeable or non-chargeable. Similarly, memory available in a node is limited.
- *Network scalability*: in case of a WSN, once the network is deployed, its extension can be done very carefully. Addition of new sensor nodes to an existing network imbalances it and hence affects the network performance.
- *Distance management*: sensor node operation is managed remotely. Numerous sensor nodes exist in a network. These sensors should work in unison in a controlled environment to meet the purpose. Again, in case of partial node failure, there is the requirement of reconfiguration of network which can be initiated from the base station, which is placed very far from the sensor nodes. Issues like message loss, data fusion, and data transmission must be handled properly.
- *Time-constraint data*: some data in a real-time environment are generated with respect to time, and this data must be delivered to the base station immediately for taking timely action. Timely delivery of the real-time data is a requirement, and suitable protocols should be designed to achieve the goal. An issue like network congestion also requires attention.
- *Node density*: this is the number of nodes deployed in the sensor network. A WSN may be dense or sparse, depending on the application. Deciding the exact number of sensor nodes required for smooth operation of the sensor network is a difficult task. We cannot randomly choose a figure because it is related to cost factor as well as performance factor. There should be a balance between cost and performance.
- *Transmission range*: the actual transmission range of a sensor node is not achieved practically due to certain environmental factors like weather, humidity, terrain, pressure, etc. So, routing protocols should have the intelligence to handle this type of situation.
- *Message loss*: this occurs mainly due to congestion in the network. Other reasons are buffer overflow, death of neighboring nodes present within the communication range of a sensor node, etc.
- *Network coverage*: one of the major deciding factors in measuring the performance of WSN is network coverage. If the sensor nodes are not properly deployed, there might be some region in the monitoring area with no sensor and hence remain unattended. This unattended area is called 'hole'. There might be more than one hole in the network. Presence of multiple holes in the network is a constraint in meeting the purpose of using the network.

- *Connectivity*: the nodes in the network work collectively to do the task for which it is designed. But sometimes it is seen that the node-to-node communication fails due to communication link failure. The reason might be hardware failure, software failure, etc. Frequent link failure interrupts the service consumers. So, there should be robust hardware and software to handle the situation.
- *Security*: there are the chances of tampering of sensor nodes deployed in very remote areas. The following risks are there with the WSN:
  o Possibility of interception in node-to-node communications or node-to-base station communication and modification in the message
  o Possibility of data theft
  o Possibility in changing the communication protocols

3.3   FACTORS DECIDING WSN LIFETIME

Wireless sensor networks are battery powered and are with limited storage space and computational power. The power supply of the network greatly affects the operation of the sensor devices and hence the network operation. In order to get prolonged service from the deployed network, it is essential to have smooth management of the available resources, i.e., energy and memory space. Different factors that decides WSN lifetime are (Figure 3.3):

- *Type of network.*
  o *Static vs. dynamic network.* In case of static WSN, once the network is deployed, network configuration changes only when network partitions into different parts due to node failure or when congestion occurs. Location of neither sensor nodes nor base station changes. But in the case of dynamic WSN, both sensor nodes and sink are mobile. Sometimes, only the sink is mobile. So, there is the frequent requirement of network reconfiguration with the mobility of the network. Of course, the network may contain a mixture of static and mobile nodes. Frequent reconfiguration consumes more battery power.
  o *Homogeneous vs. heterogeneous network.* In case of the homogeneous sensor network, all the sensor nodes have similar properties including same battery power, whereas in the heterogeneous sensor network, sensor nodes are with varying battery power, varying sensing range, and varying communication range. Heterogeneous sensor networks are better than homogeneous sensor networks in terms of network cost, maintenance cost, and better power utilization.
- *Type of battery used.*
  Depending upon the application, sensors with chargeable batteries or non-chargeable batteries are used. The battery discharge characteristic depends on the chemical used in the battery.
  o *Chargeable battery.* Lithium polymer batteries are chargeable. The lifetime of sensor network consisting of sensors containing chargeable batteries are more in comparison to sensor network consisting of sensors with non-chargeable batteries.

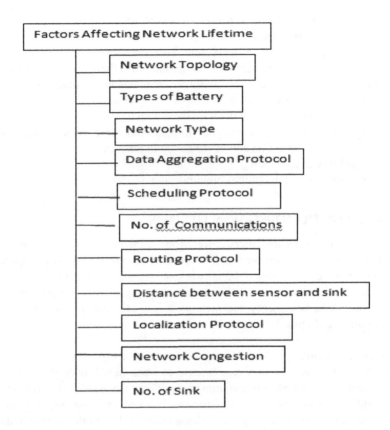

FIGURE 3.3 Factors affecting network lifetime.

- o *Non-chargeable battery.* Alkaline, nickel cadmium, and nickel metal hydride are non-chargeable batteries.
- *Network topology.* Different sensor network topologies used are grid topology, mesh topology, bus topology, tree topology, circular topology, ring topology, and star topology. There is varied energy consumption in different topologies. This depends on how much overhead information is generated in each topology.
- *Data aggregation protocol.* Sensor-generated data are aggregated and undergoes in-node processing before it is sent to the sink using multi-hop communication. At the time of processing, redundant data are eliminated from the collected data and compressed to reduce the size of data. The objective of the compressing data is to save the network bandwidth and hence to reduce overall power consumption. Different approaches for data aggregation are used: centralized approach, in-network approach, tree-based approach, and cluster-based approach. Energy requirement for data aggregation depends on the type of aggregation method used.
- *Amount of communication in the network per unit time.* When the amount of node-to-node communication is greater, power consumption is greater. Some

of the reasons for increased amount of communication are network congestion, small size packet, more overhead information generation, retransmission of packet in case of packet loss, etc.

- *Scheduling protocol.* To have collision-free communication in WSN, time slots must be assigned to the sensor nodes so that they can send/receive packets in their allotted slot. This allows periodic real-time flow of data from one node to another in a synchronized way. Use of proper scheduling protocol minimizes the amount of message exchange and hence the power consumption. Depending upon the type of WSN, it is decided which protocol will be used for this purpose. Some of the scheduling protocols used in WSN are EEWS, MDS, DAS,TRAMA, DMAC, etc.
- *Routing protocol.* Routing protocols are used to determine path of transmission of packets from sender to receiver. Always, shorter and congestion free path is preferred for packet transmission in order to avoid delay in transmission. Routing tables are maintained at the nodes which maintain path information. At the time of packet transmission, corresponding path in between two nodes are searched. If it is present, it is retrieved and used for transmission. If not present, a new path is recorded in the routing table. A small-size routing table occupies less memory space and also time of searching relevant path is less and hence saves energy.
- *Distance between sensor node and base station.* More overhead information is generated when the distance between sensor node and base station is greater. This is because the packet from the sender reaches the receiver's end through multi-hop communication, and the path information is added to the packet at each hop. Wastage of energy occurs while processing this overhead information.
- *Localization protocol.* To facilitate node-to-node communication, location information of both sender and receiver must be known. Localization protocols are used for this purpose. Different types of node localization protocols are used in WSNs. There should be a tradeoff between location information and energy efficiency, as both are desired at the same time. Volume of message exchange at the time of node localization varies from protocol to protocol and hence the energy consumption at the time of localization does as well.
- *Network congestion.* If the entire WSN or a part of the WSN is congested by traffic, packet transmission is blocked. After a fixed period of time, the network revives and retransmission of same packets is done again, causing unnecessary consumption of energy.
- *Presence of multiple sink.* Presence of multiple sink in WSN achieves faster computation and reliability of the network. At the same time, it increases the complexity of the system which leads to energy waste.

3.4 ARTIFICIAL INTELLIGENCE TECHNIQUE

Artificial intelligence (AI) is a process of simulating human intelligence by machines to solve real-world problem. Smart machines with human-type analyzing ability are built to solve complex problems in less time and with greater accuracy. The machine learning (ML) technique is a subset of AI, and the deep learning (DL) technique is a subset of ML. Machine learning applies statistical technique for data processing, whereas deep learning is a type of machine learning that runs input through a biologically inspired neural network. Not all AI techniques are ML techniques, but the reverse is true. Similarly, not all DL techniques are ML techniques but the reverse is true. Generally, AI is divided into three categories: narrow AI, artificial general intelligence (AGI), and artificial superintelligence (ASI) [8].

- *Narrow AI*, also known as weak AI, operates within a limited context and is a simulation of human intelligence. It focuses on performing a single task extremely well, and while these machines seem intelligent, they are operating far more constraints and limitations than even the most basic human intelligence. This is the most successful realization of AI today. Some examples of narrow AI are Google Search, image processing software, self-driving cars, personal assistants, and IBM's Watson. Machine learning and deep learning belong to this category.
- *Artificial general intelligence*, also referred as strong AI, is a kind of intelligence similar to human intelligence and can be used to solve any problem. It has the ability of self-learning for self-improvement. We see this type of intelligence in robots.
- *Artificial superintelligence* is an aspect of intelligence which is more powerful and sophisticated than a human's intelligence. A human's intelligence is considered to be one of the most capable and developmental. Superintelligence can surpass human intelligence; it can think beyond what the human thinks.

Fundamental AI techniques are:

- Heuristics
- Support vector machine
- Artificial neural network
- Markov decision process
- Natural language processing
- Case-based reasoning
- Rule-based systems
- Cellular automata
- Fuzzy models
- Swarm intelligence
- Multi-agent systems
- Genetic algorithms

3.4.1 Why AI Is Required in WSN

Real-world data generated by sensors in a WSN has different unwelcome properties and hence, if used directly for processing, will lead to inaccurate result. They are

- *Voluminous data*. A large volume of continuous data are generated by sensors.

- *Unstructured data.* Collected data is unstructured data and contains some irrelevant information which need to be removed before processing
- *Varied data sources.* Different types of sensors are used for data collection.
- *Dynamic environment.* Data can be collected from a dynamic environment. For example, in a smart irrigation system, depending on the weather condition, it is decided that whether to irrigate the land or not and weather is dynamic in nature.
- *Real-time processing.* Real-time processing of collected data is required for extracting the relevant data from the generated data. This is because with the limited battery power and memory power it is not possible in the part of sensor node to store huge volume of data first and then process it.
- *Format of data.* Different approaches of data collection methods may be used by different types of sensors deployed in a network. So, format of data collection may vary from sensor to sensor. Also, data may contain missing values.
- *Non-accuracy in data.* Collected data may contain noise, and if it is true, obviously, system will generate inaccurate data. Analysis of this kind of data generates incorrect output.

It is not possible to use data with the aforementioned properties directly in a resource-constrained WSN without the help of smart techniques which can handle all these negative properties. Using AI in WSN enhances the capability of WSN to handle these properties and makes it adaptable to a dynamic environment. AI technique can be used for data aggregation, routing of packets, node deployment, etc.

3.5 WSN LIFETIME ENHANCEMENT USING AI

This section gives a brief description of the research done to achieve longer lifetime of WSN. AI can be used for different tasks in order to achieve prolonged lifetime of WSN. Some of them are data aggregation, network coverage and connectivity, node or object localization, scheduling, routing of packets, etc. Different approaches are used for lifetime enhancement in WSN. The approaches associated with lifetime enhancement are energy harvesting, energy transfer/charging, and energy conservation [5]. Energy harvesting is a mechanism used by sensors to generate energy from the ambient surrounding to provide uninterrupted power supply and to save the network from untimely death. Energy transfer is the process of transferring energy from energy store to the sensors in the aim of providing uninterrupted service to the user. Energy conservation is the process of consumption of available energy in the network in a controlled fashion in order to get prolonged service from the network.

3.5.1 DATA AGGREGATION USING AI

Data aggregation is the process of combining and compressing the data packets at a sensor node (called cluster head) coming from other different sensor nodes in the aim of reducing the size of data and hence minimizing the degree of packet transmission in the network. The advantage of doing this is to reduce the power consumption during data transmission and to increase the network lifetime. Some AI-based data aggregation techniques proposed by the researchers are as follows:

- An energy-efficient data aggregation scheduling process based on Q-learning is discussed in [9]. Due to the self-learning feature, the scheduling sequence automatically converges to a near-optimal sequence after a short period of exploration.
- An energy-efficient data aggregator election (EEDAE) algorithm is proposed in [10] to reduce the energy consumption, when a WSN is used to gather data. This method is used for optimal cluster node selection.
- A modified cuckoo search-based data aggregation technique is proposed in [4] for a heterogeneous network in which an attempt is made to increase the network lifetime and throughput by minimizing the amount of data.
- A novel data aggregation scheme basing on a self-organized map neural network is proposed in [11] which increases the network lifetime by reducing the redundant data and eliminating outliers. The cosine similarity is used to improve the clustering process basing on density and data similarity. All these reduce energy consumption and increase network performance.
- The neural network approach is used in [12] for data aggregation in WSN. Basically, a back propagation network and a radial basis function network are used to process data using limited energy. Before data processing, redundant data is removed from the original data.
- The ant colony algorithm-based data aggregation technique DAACA is used in [13] in which energy efficiency is achieved by minimizing the size of the routing information maintained in routing table. After a certain period of time, the routing table is updated by accommodating new route information used in the recent past for packet transmission and deleting the least used route information from the routing table.
- The genetic algorithm is used to build an energy-efficient data aggregation spanning tree, and then an acceptable route is selected which balances the network load basing upon the residual energy within the network. After the route is determined, the information from the nodes is collected by mobile agents using the artificial bee colony algorithm [14].

3.5.2 COVERAGE AND CONNECTIVITY DETERMINATION USING AI

Sensors are said to be connected if some path for communication exists among them. In a dynamic environment condition, it is not possible to have 100% continuous connectivity due to factors like change in topology, node failure, attacks on node (both software and hardware means), node isolation, network partitioning, system upgradation, and link failure. Having full connectivity and coverage in WSN are very important, as the performance of the network depends on these two factors besides other factors. The connectivity problem restricts the user in getting continuous service, whereas the coverage problem restricts the user in getting information about a particular region in the monitoring area which is unattended or isolated. This unattended area may have crucial information that could help the administrator in taking a right decision. Increasing the number of sensor nodes will help in achieving 100% coverage of the monitoring area, but it will make the system costlier and will increase the power consumption rate; hence, it is not suitable. Rate of continuous connectivity can be increased by employing monitoring software which can collect the status of devices and can take a decision automatically to restore the service. Some of the AI-based mechanisms developed so far are:

- The authors of [15] have described the connectivity issue and proposed a solution inspired by the behavior of self-organization of ants. The self-organization algorithm can be used to reconfigure sensor nodes in case of node failure, or to change the density of sensor nodes and traffic patterns.
- An intelligent black-hole algorithm with weight factor and mutation operation is used in [16] to achieve full area coverage in 3-D terrain. This algorithm is based on the concept of how a black hole devours other stars. In this algorithm, the individual with the best fitness value is regarded as a black hole, and other individuals move towards it. If the distance between an individual and the black hole is smaller than the radius of event horizon, then it will be swallowed by the black hole and an individual will be randomly generated to maintain the size of the population.
- A coverage optimization model based on improved whale algorithm is proposed in [17] for achieving full coverage in the area of interest. In this model, the idea of reverse learning is introduced into the original whale swarm optimization algorithm to optimize the initial distribution of the population. This method enhances the node search capability and speeds up the global search. Network coverage, node utilization, and energy utilization are studied.
- The particle swarm optimization (PSO) approach is used in [18] to determine the hole in the monitoring area and to fill it with additional redundant node, increasing the coverage area.
- The Nash Q-Learning-based node scheduling algorithm for coverage and connectivity maintenance (CCM-RL) is proposed in [19] where each node autonomously learns its optimal action (active/hibernate/sleep/customize the sensing range) to maximize the coverage rate and maintain network connectivity. The learning algorithm resides inside each sensor node. The main objective of this algorithm is to enable the sensor nodes to learn their optimal action so that the total number of activated nodes in each scheduling round becomes minimal and preserves the criteria of coverage rate and network connectivity.

3.5.3 Node Localization Using AI

Node localization in WSNs is an active research field in WSN because node-to-node communication is not possible without knowing the location information. Again, the location information must be accurate; otherwise, it will be of no use. Some of the existing protocols use geographical positioning system (GPS), anchor node, or beacon node to determine the location of a node. Node localization is broadly categorized into a number of categories—proximity-based localization, range-based localization, angle-and-distance-based localization, and known location-based localization. The current position of the sensor nodes in the environment may also change due to environmental calamities like cyclone, flood, etc. To handle such a situation, reconfiguration of the network may be required. Artificial intelligence can be used to improve the accuracy of location information while not affecting the network lifetime. Some research contributions of the AI-based node localization include the following:

- K-means and c-means algorithms are used in [20] to divide the monitoring field into clusters based on RSSI values and train each region to find the coordinate of a sensor node.
- Artificial Neural Network and Radial Basis Function Network are used in [21] and are used to calculate the location of unknown node. At the time of processing, RSSI information from neighboring nodes is used.
- A device-free wireless localization system using artificial neural networks (ANNs) is proposed in [22]. The system consists of two phases. In the offline training phase, Received Signal Strength (RSS) difference matrices between the RSS matrices collected when the monitoring area is vacant and the input to the ANN is calculated and training phase starts. In the localization phase, coordinates of sensors are calculated.
- In [23], a node localization scheme is proposed based on a recent bio-inspired algorithm called the Salp Swarm Algorithm (SSA). The proposed algorithm is compared to well-known optimization algorithms, namely, particle swarm optimization (PSO), the butterfly optimization algorithm (BOA), the firefly algorithm (FA), and grey wolf optimizer (GWO), under different WSN deployments. It is found that SSA is better in comparison in terms of accuracy in localization information.
- A metaheuristic algorithm named bat algorithm is used in [24], where node localization problem is viewed as an optimization problem in multidimensional space in large WSN. Localization of nodes is based on the velocity of bats, i.e. the nodes, and the Doppler Effect. Doppler Effect is used to minimize the localization error at the time of bat movement.
- The butterfly optimization algorithm is proposed in [25] which is used for node location estimation from the distance information between nodes which are corrupted by Gaussian noise.

3.5.4 ROUTING USING AI

Sensor nodes communicate with each other through multi-hop communication. Path finding for a packet at the time of routing is a complex task because a lot of factors are considered, like network traffic, transmission range, residual energy of nodes, network size, terrain, packet size, packet rate, current state of receiver, and distance between sender and receiver. All these require a lot of information processing, which consumes a lot of energy. The artificial intelligence technique can be used to improve routing and hence to improve the network performance including lifetime of the network. Some of the techniques developed by the researchers are:

- A new artificial intelligence-based routing technique called 'Sensor Intelligence Routing' (SIR) is introduced in [26]. This protocol uses artificial neural network (ANN) with self-organizing map (SOM) where every neuron is treated as a sensor and the connection between nodes represent the route. Each connection is a weighted connection which indicates the distance between the nodes or the network traffic in between two nodes. The weight on the connection is updated gradually when traffic flows over a link. The best path for a packet is chosen basing upon the current traffic in the network.
- A model based on density-based spatial clustering of application with noise (DBSCAN) is suggested in [1], in which a spatiotemporal relational model of sensor nodes is built. With the application of nature-inspired algorithms like ant colony optimization (ACO), bees colony optimization (BCO), and simulated annealing (SA), the optimal path for packet transmission is determined.
- A hybrid Power Efficient Gathering in Sensor Information System (PEGASIS) hierarchical protocol is proposed in [27]. This protocol uses the firefly optimization technique and artificial neural network for enhancing the lifespan of the WSN. The sensor node location is random, and every node has the capability to detect data, blend data, and equally send the load among the nodes. A chain of nodes is made according to the positioning of the node, and the nodes are plotted by using the greedy algorithm. Artificial neural network (ANN) is used as a classifier to remove distortion from the network or to overcome the battery discharge problem. The firefly algorithm is used to arrange the sensor nodes in grid. The conventional routing protocols based on computational intelligence techniques have some drawbacks, viz., slow convergence rate, large memory constraints, high sensitivity to initial value, large communication overhead, and high learning period.
- The opportunistic routing (OR) protocol is one of the new routing protocol that promises reliability and energy efficiency during transmission of packets in WSNs is discussed in [28]. An intelligent opportunistic routing protocol (IOP) using a machine learning technique is selected, to select a relay node from the list of potential forwarder nodes to achieve energy efficiency and reliability in the network.
- An attempt is made to increase the lifetime of the network by controlling the node mobility basing upon the residual energy in [29]. Taking residual energy of nodes into consideration, a graph is constructed and shortest path is determined for packet transfer from sender to receiver.
- In [30], the cross layer approach with constrained least square method is used to yield an analytical solution of the optimal network flow at each link in order to maximize the network lifetime in case of underwater wireless sensor network.
- Fuzzy logic based unequal clustering and ACO-based routing is used in [31] for cluster head selection, inter-cluster routing, and cluster maintenance and to enhance the network lifetime.

- In [32], a delay-tolerant WSN is taken into consideration where the sensor node is not bound to immediately send the collected information to the mobile sink within a threshold period. Rather, they wait till the mobile sink becomes nearer to them and hence avoids long-distance transmission and saves energy.
- The shuffled frog leaping algorithm (SFLA) is used in [33] for routing of packets for having balanced energy consumption in the network. Building a new energy consumption model, minimizing long range signal characteristics, and increasing network lifetime are the key objectives achieved.

3.5.5   SCHEDULING USING AI

In order to have smooth communication among sensor nodes and to avoid collision among packets, which results in packet loss, it is required to assign time slots to nodes for communication. The absence of a proper scheduling algorithm causes retransmission of packets, which is one of the reasons for unnecessary consumption of energy and thereby limits the network lifetime. With the growing application of artificial intelligence in different fields, the researchers have also tried it for time scheduling. Some of the researches done in this regard are described as follows:

- Sensing range adjustable nodes are used in the network considered in [34] for having varying energy consumption rate, and the neighborhood-based distribution algorithm (NEDA) is used to activate the sensors at a particular slot to monitor the surroundings. Then, the linear programming-based technique is used to assign the activation time.
- Scheduling of mobile sink movement is done in [35]. The lifetime of the network is increased by balancing the sink movement using the hyperheuristic framework that can schedule sink movement automatically in both static and dynamic environments and can avoid unnecessary control overhead generation.
- The communication weighted greedy cover (GWGC) algorithm is used in [36] to schedule the sensor node partitioning into multiple sets in which both coverage and connectivity among sensor nodes and sink is achieved separately, and then to build a maximum tree cover to maintain coverage and connectivity while maintaining the network lifetime.
- An energy-efficient, cluster-based scheduling algorithm is proposed in [37] in which after creation of cluster of nodes, the cluster head is chosen basing on the energy currently available with the nodes. Then, a time slot is allocated to each cluster for packet transmission. The author proposed the energy consumption model as well as a new packet format.

3.5.6   NODE DEPLOYMENT USING AI

The deployment of sensor nodes in monitoring area is one of the major concerns in WSN because its performance depends highly on the position of sensor nodes. Key design objectives in deploying sensor nodes include coverage, energy consumption, a lifetime of a network, connectivity, and cost, as represented by the number of sensor nodes. The Quality of Service (QoS) and the deployment cost as well as maintenance cost of the network are intercorrelated. The optimal sensor node placement enables the administrator to minimize manpower and time to acquire accurate information on the monitoring area. Energy consumption, coverage, connectivity, and cost of deployment are conflicting objectives. The cost of deployment can be minimized by minimizing the sensor nodes. But this may be a problem in achieving high connectivity and coverage in network. It may also be a reason for increasing energy consumption as the distance between nodes increases. So, there should be a balance of energy consumption, high coverage and connectivity, and cost of deployment. AI techniques can be used to determine the proper node density so that all the objectives of the network can be achieved in a balanced way. Also, it can help to determine the optimized position of the sensor nodes [6]. Node deployment algorithms can be divided into two broad categories: static and dynamic. In the static node deployment scheme, position of the nodes remains unchanged throughout the entire network lifetime once they are deployed; whereas with dynamic deployment, the node position can be changed. One of the major advantages of dynamic deployment is that the hole can be minimized by moving the sensor node to the unattended portion of the monitoring area. Static deployment is used in case of homogeneous network, and dynamic deployment is used in case of both homogeneous and heterogeneous network. Some of the research work done in this regard is as follows:

- In the evolutionary approach-based Voronoi Diagram (EAVD), stationary sensor nodes are randomly deployed, then the area is divided into Voronoi cells. The genetic algorithm (GA) is then used to deploy additional mobile sensor nodes in each cell to heal coverage holes [38].
- A genetic algorithm and Voronoi Vertex averaging algorithm (VVAA) is proposed in [39] to relocate the mobile sensor nodes in addressing the coverage holes problem. GA is used to find the optimum locations, while coverage holes are detected by using the Voronoi Diagram. The algorithm offers the highest throughput. Here, network coverage is addressed while ignoring all other parameters.
- The glowworm swarm optimization (GSO)-based node placement strategy is used in [40]. Here, an attempt is made to maximize the coverage for a fixed number of mobile sensor nodes. This work is inspired by the behavior of a glowworm that carries a luminescent substance called luciferin. The movement of a glowworm is decided by the intensity of luciferin possessed by its neighbors. Each glowworm will be attracted towards the brighter glow of other glowworms in the neighborhood, and it will move towards the brightest neighbor. Each sensor node is treated as an individual glowworm and moves towards its neighbor that has lower intensity of luciferin. The intensity of the luciferin decreases with distance. A sensor node receives luminance from other sensor nodes if it is present within its communication range. This algorithm focuses on coverage and energy consumption.
- The artificial bee colony (ABC) optimization technique is used for sensor node deployment in [41]. ABC was developed based on the foraging behavior of a honey bee swarm. The objective of this work is to rearrange mobile sensor

nodes in a homogeneous WSN that maximizes the coverage rate of the network.
- The territorial predator scent marking algorithm (TPSMA) to redeploy mobile sensor nodes is used in [42]. The algorithm imitates the behavior of a territorial predator in marking their territories with their odors. Focus is on maximum coverage and minimum energy consumption, connectivity, and network cost.
- The ACO technique is used in [42] to optimize the number of sensor nodes while the constraints are full coverage and full connectivity. Initially, a random number of sensor nodes are deployed which are then minimized with the help of the ACO technique.

3.6   CONCLUSION

Lifetime is a key factor in a resource-constrained WSN. As long as the nodes are active, we can get uninterrupted service from the network. But when some of the nodes or all of the nodes starts to fail, the operation of the network is interrupted. WSN has a wide range of application starting from health applications to education. So, a prolonged network service is desired from it. In WSN, energy consumption occurs at the time of data generation, processing and transmission/retransmission of packets, node/object localization, data aggregation, network coverage and connectivity determination, etc. Protocols used to perform all these operations in computer network cannot be used for WSN as sensors are with limited storage, limited processing capability, and limited battery power. So, there is the requirement of developing the protocols that are well suited for these constraints. In this chapter, brief descriptions of issues related to WSN were discussed, as were the constraints in achieving prolonged network lifetime. Then a brief description was given of researches done by different researchers so far. To increase the lifetime of the network, it is required to minimize the energy consumption in the network after its deployment in the monitored area. One of the ways to minimize the energy consumption is to perform a network operation in a smarter way which will save time and generate less control overhead information. Artificial intelligence is an alternative way of doing the tasks in a better manner. Embedding AI with WSN makes the system intelligent and makes it adaptive to a dynamic real-time environment. In this chapter, it was described how artificial intelligence technique can be used in data collection and aggregation, node deployment, routing, node localization, network coverage and connectivity determination, scheduling. etc. From the researches it is evident that the WSN lifetime can be increased with the application of AI to it.

REFERENCES

1. Srinivas Narsegouda M., Umme Salma, Anuradha N. Patil, "Nature Inspired Algorithm Approach for the Development of an Energy Aware Model for Sensor Network," *SpringerLink, Computational Intelligence in Sensor Networks*, May 2018, 55–77.
2. Parulpreet Singh, Khosla Arun, Kumar Anil, Khosla Mamta. "Computational Intelligence Techniques for Localization in Static and Dynamic Wireless Sensor Networks-A Review," *SpringerLink, Computational Intelligence in Sensor Networks*, May 2018, 25–54.

3. Sahoo Biswa Mohan, Tarachand Amgoth, Hari Mohan Pandey. "Particle Swarm Optimization Based Energy Efficient Clustering and Sink Mobility in Heterogeneous Wireless Sensor Network." *Ad Hoc Networks*, 106, 102237, 2020.
4. Sahoo Biswa Mohan, Hari Mohan Pandey, Tarachand Amgoth. "GAPSO-H: A Hybrid Approach Towards Optimizing the Cluster-Based Routing in Wireless Sensor Network." *Swarm and Evolutionary Computation*, 60, 100772, 2021.
5. Engmann Felicia, Ferdinand Apietu Katsriku, Jamal-Deen Abdulai, Kofi Sarpong Adu-Manu, Frank Kataka Banaseka, "Prolonging the Lifetime of Wireless Sensor Networks: A Review of Current Techniques," *Hindawi Wireless Communications and Mobile Computing*, 2018, Article ID 8035065.
6. Abidin Husna Zainol, Norashidah Md. Din, Nurul Asyikin Mohamed Radzi, Zairi Ismael Rizman, "A Review on Sensor Node Placement Techniques in Wireless Sensor Networks," *International Journal on Advanced Science, Engineering, Information Technology*, 7, 1, 2017, ISSN: 2088-5334.
7. Indu Sunita Dixit, "Wireless Sensor Networks: Issues & Challenges," *International Journal of Computer Science and Mobile Computing*, 3, 6, 681–685, June 2014, ISSN 2320-088X.
8. Panchal Shubham, "Types of Artificial Intelligence and Examples," https://medium.com/predict/types-of-artificial-intelligence-and-examples-4f586489c5de.
9. Lu Yao, Taihua Zhang, Erbao He, Ioan-Sorin Comsa. "Self-Learning-Based Data Aggregation Scheduling Policy in Wireless Sensor Networks," *Hindawi Journal of Sensors*, 2018, Article ID 9647593, https://doi.org/10.1155/2018/9647593.
10. Subashree C. P., S. Thangalakshmi," Energy Efficient Aggregation In Wireless Sensor Networks Using Artificial Intelligence Based Aggregator Election," *International Refereed Journal of Engineering and Science (IRJES)*, ISSN (Online) 2319-183X, (Print) 2319-1821 5, 3, 8–16, (March 2016).
11. Ullah Ihsan, Hee Yong Youn, "A Novel Data Aggregation Scheme Based On Self-Organized Map for WSN," *The Journal of Supercomputing*, 75, 7, July 2019. https://doi.org/10.1007/s11227-018-2642-9.
12. Khorasani Fereshteh, Hamid Reza Naji, "Energy Efficient Data Aggregation in Wireless Sensor Networks Using Neural Networks," *Inderscience, International Journal of Sensor Networks*, 24, 1, 26–42, 2017.
13. Lin Chi, Guowei Wu, Feng Xia, Mingchu Li, Lin Yao, Zhongyi Pei, "Energy Efficient Ant Colony Algorithms for Data Aggregation in Wireless Sensor Networks," *Journal of Computer and System Sciences*, 78, 6, 1686–1702, November 2012.
14. Thangaraj M. and P. P. Ponmalar, "*Swarm intelligence based secured data aggregation in wireless sensor networks, 2014 IEEE International Conference on Computational Intelligence and Computing Research*, Coimbatore, 2014, pp. 1–5. https://doi.org/10.1109/ICCIC.2014.7238519.
15. Benahmed Khelifa H., Haffa Merabti Madjid, David Llewellyn-Jones, "Monitoring Connectivity in Wireless Sensor Networks," *International Symposium on Computers and Communications*, July 2009.
16. Pan Jeng-Shyang et al. "3-D Terrain Node Coverage of Wireless Sensor Network Using Enhanced Black Hole Algorithm." *Sensors (Basel, Switzerland)*, 20, 8 2411.23 Apr. 2020. https://doi.org/10.3390/s20082411.
17. Wang Lei, Weihua Wu, Junyan Qi, Zongpu Jia, "Wireless Sensor Network Coverage Optimization Based on Whale Group Algorithm," *Computer Science and Information Systems*, 15(3):569–583, 2018. https://doi.org/10.2298/CSIS180103023W.
18. Mehta Shalu, Amita Malik," A Swarm Intelligence Based Coverage Hole Healing Approach for Wireless Sensor Networks," *EAI Endorsed Transactions on Scalable Information Systems*, February 2020. https://doi.org/10.4108/eai.13-7-2018.163132.

19. Sharma Anamika, Siddhartha Chauhan, "A Distributed Reinforcement Learning Based Sensor Node Scheduling Algorithm for Coverage and Connectivity Maintenance in Wireless Sensor Network," *SpringerProfessional, Wireless Network*, 26, 6: 4411–4429, 2020.

20. Bernas M., B. Placzek, "Fully Connected Neural Networks Ensemble With Signal Strength Clustering for Indoor Localization in Wireless Sensor Networks," *International Journal of Distributed Sensor Network*, 11(12), 1–10, 2015.

21. Madagouda B. K., R. Sumathi, *"Analysis of Localization Using ANN Models in Wireless Sensor Networks," 2019 IEEE Pune Section International Conference (PuneCon)*, Pune, India, 2019, pp. 1–4. https://doi.org/10.1109/PuneCon46936.2019.9105871.

22. Sun Yongliang, Xuzhao Zhang, Xiocheng Wang, Xinggan Zhang, "Device-Free Wireless Localization Using Artificial Neural Network in Wireless Sensor Networks," *Wireless Communications and Mobile Computing*, 2018,Article ID 4201367. https://doi.org/10.1155/2018/4201367.

23. Kanoosh Huthaifa M., Essam Halim Houssein, Mazen M. Selim, "Salp Swarm Algorithm for Node Localization in Wireless Sensor Networks," *Hindawi Journal of Computer Networks and Communications*, 2019, Article ID 1028723. https://doi.org/10.1155/2019/1028723.

24. Mihoubi Miloud, Abdellatif Rahmoun, Pascal Lorenz, Noureddine Lasla," An Effective Bat Algorithm for Node Localization in Distributed Wireless Sensor Network," wileyonlinelibrary.com/journal/spy 2,2017. https://doi.org/10.1002/spy2.7

25. Arora Sankalp, Satvir Singh, "Node Localization in Wireless Sensor Networks Using Butterfly Optimization Algorithm," *SpringerLink, The Arabian Journal for Science and Engineering*, 42, 3325–3335, 2017. https://doi.org/10.1007/s13369-017-2471-9.

26. Barbancho J., Leon C., Molina J., Barbancho A. (2006), "SIR: A New Wireless Sensor Network Routing Protocol Based on Artificial Intelligence." In: Shen H.T., Li J., Li M., Ni J., Wang W. (eds.) *Advanced Web and Network Technologies, and Applications*. APWeb 2006. Lecture Notes in Computer Science, vol 3842, Springer, Berlin, Heidelberg. https://doi.org/10.1007/11610496_35

27. Ali S. and R. Kumar, *"Artificial Intelligence Based Energy Efficient Grid PEGASIS Routing Protocol in WSN," 2018 7th International Conference on Reliability, Infocom Technologies and Optimization (Trends and Future Directions) (ICRITO)*, Noida, India, 2018, pp. 1–7. https://doi.org/10.1109/ICRITO.2018.8748501.

28. Bangotra Deep Kumar, Yashwant Singh, Arvind Selwal, Nagesh Kumar, Pradeep Kumar Singh, Wei Chiang Hong, "An Intelligent Opportunistic Routing Algorithm for Wireless Sensor Networks and Its Application Towards e-Healthcare," *Sensors*, 20(14), 3887, 2020. https://doi.org/10.3390/s20143887.

29. Mahboubi Hamid, Walid Masoudi Mansour, Amir G. Aghdam, Kamran Sayrafian-Pour, "Maximum Lifetime Strategy for Target Monitoring With Controlled Node Mobility in Sensor Networks With Obstacles," *IEEE Transactions on Automatic Control*, 61, 11, 3493–3508, November 2016.

30. Zhou Yuan, Hongyu Yang, Yu-Hen Hu, Sun-Yuan Kung, "Cross-Layer Network Lifetime Maximization in Underwater Wireless Sensor Networks," *IEEE Systems Journal*, 14, 1, 220–231, March 2020.

31. Arjunan Sariga, Pothula Sujatha, "Lifetime Maximization of Wireless Sensor Network Using Fuzzy Based Unequal Clustering and ACO Based Routing Hybrid Pprotocol," *Applied Intelligence*, August 2018. https://doi.org/10.1007/s10489-017-1077-y.

32. Yun Young Sang, Ye Xia, "Maximizing the Lifetime of Wireless Sensor Networks with Mobile Sink in Delay-Tolerant Applications," *IEEE Transactions on Mobile Computing*, September 2010. https://doi.org/10.1109/TMC.2010.76.

33. Zhou Chunliang, Ming Wang, Weiqing Qu, Zhengqiu Lu, "A Wireless Sensor Network Model considering Energy Consumption Balance," *Hindawi Mathematical Problems in Engineering*, 2018, Article ID 8592821. https://doi.org/10.1155/2018/8592821.

34. Chen Zong-Gan, Ying Lin, Member, Yue-Jiao Gong, Zhi-Hui Zhan, Jun Zhang, "Maximizing Lifetime of Range-Adjustable Wireless Sensor Networks: A Neighborhood-Based Estimation of Distribution Algorithm," *IEEE Transactions on Cybernetics*, February, 2020. https://doi.org/10.1109/TCYB.2020.2977858.

35. Zhong J. H., Z. X. Huang, L. Feng, W. Du, and Y. Li, "A Hyperheuristic Framework for Lifetime Maximization in Wireless Sensor Networks with a Mobile Sink," *IEEE/CAA Journal of Automatica Sinica*, 7, 1, 223–236, Jan. 2020.

36. Zhao Qun, Mohan Guruswamy, "Lifetime Maximization for Connected Target Coverage in Wireless Sensor Networks," *IEEE/ACM Transactions on Networking*, December 2008. https://doi.org/10.1109/TNET.2007.911432.

37. Janani E. Srie Vidhya and P. Ganesh Kumar, "Energy Efficient Cluster Based Scheduling Scheme for Wireless Sensor Networks," *The Scientific World Journal*, 2015, Article ID 185198. http://dx.doi.org/10.1155/2015/185198.

38. Sengupta S., S. Das, M. D. Nasir, and B. K. Panigrahi, "Multiobjective Node Deployment in Wsns: In Search of an Optimal Tradeoff Among Coverage, Lifetime, Energy Consumption, and Connectivity," *Engineering Applications of Artificial Intelligence*, 26, 405–416, Jan. 2013.

39. Juli V.V. and J. Raja, "Mobility Assisted Optimization Algorithms for Sensor Node Deployment," *European Journal of Scientific Research*, 78, 156–167, 2012.

40. Liao W. H., Y. C. Kao, and Y. S. Li, "A Sensor Deployment Approach Using Glowworm Swarm Optimization Algorithm in Wireless Sensor Networks," *Expert Systems with Applications*, 38, 12180–12188, Sep. 2011.

41. Ozturk C., D. Karaboga, and B. Gorkemli, "Artificial Bee Colony Algorithm for Dynamic Deployment of Wireless Sensor Networks," *Turkish Journal of Electrical Engineering and Computer Sciences*, 20, 255–262, Feb. 2012.

42. Fidanova S., P. Marinov, and E. Alba, "Ant Algorithm for Optimal Sensor Deployment," *Computational Intelligence*, 399, 21–29, 2012.

4 Research Issues of Information Security Using Blockchain Technique in Multiple Media WSNs
A Communication Technique Perceptive

Nihar Ranjan Pradhan and Akhilendra Pratap Singh

National Institute of Technology, Meghalaya, India

CONTENTS

DOI: 10.1201/9781003145028-4

4.1 INTRODUCTION: BACKGROUND AND DRIVING FORCES

The wireless sensor network (WSN) in conventional and non-conventional media has numerous applications, such as hospital records, tracking and connectivity in automobiles (VANET), detecting weather conditions, monitoring soil condition in agriculture, earthquake and landslides, pipeline leakage detection, surveillance in defense, body area networks, identification of animals using bio-chips, etc. MI-based communication is widely used where electromagnetic (EM), acoustic, and optical communication fail [3]. Communication such as Zig-Bee, Wi-Fi, and Bluetooth has a lesser range. However, the data collected at sensor nodes are risky. The data collected may be breached, tampered, or forged.

WSN is a centralized approach and hence is vulnerable to attacks, whereas Blockchain technology is a decentralized and distributed architecture which can be implemented for data security of WSN devices with consensus mechanism and digital signature [4]. Blockchain technology combines the advantages of a decentralized, distributed ledger; reduced transaction cost; faster transaction settlement; transparency; cryptographic property; enhanced security; and immutability to ensure the communication [2,3]. Sensors interconnect among themselves through private and public networks and can be controlled remotely [5]. The communication between sensors takes place through some standard protocol and centralized approach. Mostly, these sensors are heterogeneous in nature, with limited power and memory, so security solutions need to be adapted 3]. We have to be careful about resource consumption while proposing new solutions for any problem.

WSN devices utilize enough recourse to support the consensus algorithm such as proof of work (PoW) used in Blockchain. So, an ideal approach involves not using PoW. Local Blockchain is centrally managed by the owner or cluster head. In recent years, tremendous effort has been made for various approaches targeting security issues and threats. Key generation and management is the most significant aspect of security, such as privacy and authentication. So, the aim is to keep these streams of bits confidential by applying some complex mathematical operations. Performing only mathematical operations cannot secure data and information. As security is the major feature in a WSN, we propose implementing Blockchain in it. A Blockchain technology is a distributed public ledger of all transactions among nodes. Each verified transaction is accumulated in a block. Each block consists of a variable number of verified transactions. The consensus technique is to confirm that the source node is trusted rather than a malicious node. Blockchain uses the cryptographic method to send secret information so that the target node can decrypt the information (Figure 4.1). The main contributions of our work are highlighted as follows:

a. The bibliometric review of papers has been carried out. Some interesting application and advantages of Blockchain-based WSN have been reported in the literature.

b. The Blockchain architecture, characteristics, consensus algorithms, cryptocurrency, smart contracts, and use cases are systematically discussed. Classification, advantages, and disadvantages of Blockchain are also reported in the literature. The cryptographic principles of Blockchain technology, such as SHA, ECC, digital signature, and Merkle tree are included. The fundamental motivation

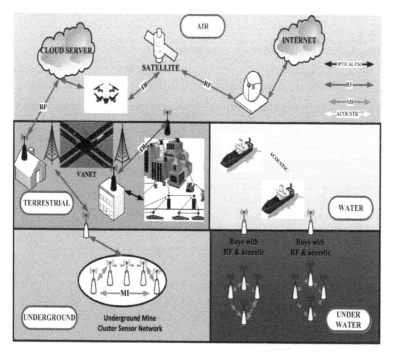

FIGURE 4.1 Blockchain-based WSN.

behind this segment is to tell individuals about the basics of Blockchain technology.

c. This chapter introduces the idea of security threats in WSN and how communication links, nodes, and data can be secured with the help of Blockchain.

d. Finally, this chapter describes the list of projects and funding agencies worldwide, and how developing countries implement Blockchain-based services.

4.2 BACKGROUND AND MOTIVATION

The rapid growth in Blockchain technology in recent years has motivated us to conduct the research studies and challenges from a different perspective. It has also opened many directions and research gaps for research scholars. Blockchain application has revolutionized various sectors, including IoT [6–9], supply chain management [10,11], Healthcare [12,13], finance [14], and government applications [15,16]. So, we aim to conduct a comprehensive survey on existing Blockchain applications, current state of art in this direction and also tried to find out the missing link between WSN communication (RF and MI) and Blockchain security. In this regard, a bibliometric analysis [17] is done within the related domain of Blockchain. All the Blockchain and related conference papers, journals, survey papers, and articles have been indexed by the existing scientific database SCOPUS and Google Scholar from 2013 to 2019, as the first paper on Blockchain was published in 2013. The result also gives yearly publication in Table 4.1, a list of supportive project funding bodies from industry and academics, and a list of top 10 cited Blockchain papers in Table 4.2. After choosing SCOPUS

TABLE 4.1
List of Blockchain Application in Year Wise

SL No	Blockchain Application	Country	Implementation Year
1	French court for corporate registry [16]	France	2019
2	Postal services	Italy	2019
3	Spanish port authority	Spain	2019
4	Air traffic management by NASA	US	2018
5	Hello Tractor for agriculture	Africa	2018
6	Trading	European Union	2018
7	Supply chain logistic network, US Air Force	US	2018
8	Blockchain Island Cryptocurrency	Malta	2018
9	E-residency program	Estonia	2017
10	Digital ID based on Ethereum	Switzerland	2018
11	Smart Dubai office, SDO	UAE	2016
12	Blockchain University	UK	2016
13	Blockchain-Based property title	Sweden	2016

TABLE 4.2
List of Top Cited Blockchain

Topic	2013	2014	2015	2016	2017	2018	2019	Total
Blockchain	2	9	37	174	780	2394	1424	4820
Bitcoin	16	72	143	176	273	262	95	1037
Cryptocurrency	0	10	31	52	93	117	295	598
Ethereum	0	0	3	13	47	57	125	245
Smart contract	1	0	7	34	120	108	141	411
WSN Security	1106	1033	1168	1210	1197	1297	492	7503
WSN	9253	9120	8816	8776	8761	9160	3396	57282

and Google Scholar as the scientific search engine we found some related terminology such as Blockchain [18–21], Cryptocurrency, Smart contract [22], Ethereum [23], Bitcoin & WSN security [24] as query string to extract the papers. Table 4.2 illustrates that the research carried out in Blockchain area grows more and more through years. The search was done on June 11, 2019. The result in Table 4.3 shows the significance of Blockchain based WSN and motivated us to choose Blockchain [17].

4.3   APPLICATION OF BLOCKCHAIN-BASED WSN COMMUNICATION

4.3.1   RFID-BASED FOOD SUPPLY CHAIN

Food safety has become an important problem worldwide. As we know, many foods are perishable and need certain temperature and humidity for preservation. In developed countries, the loss percentage of agrifoods is 3%, whereas in developing and

TABLE 4.3
Overview of Significance of Blockchain-Based WSN

Communication Techniques	Advantages	Disadvantages	Application WRT Media
Acoustic	Long range up to 20 KM. Long haul communication. Most widely used underwater communication technology.	Costly and energy consuming trans receivers. Harmful to some marine life. Large communication latency. Data transmission rate is low.	Air, terrestrial, underground, underwater, military submarines or surveillance
RF	Data transmission rate is medium at close distance. Suitable for air and shallow waters, short distance local underwater communication. Tolerant power is more for water turbulence and turbidity. Smooth transaction to cross air/water boundaries.	It has a short link. Costly and energy consuming	Air-satellite communication terrestrial-smart cities, VANET, body area network underground-underwater
Optical [33]	Low cost and small volume transmission. Immune to transmission latency. Ultra-high data transmission rate.	Moderate link range. It suffers from severe absorption and scattering. It can't cross water/air boundaries.	Air-terrestrial-mobile communication; Underground-mobile communication underwater
FSO	Can be used for ultra-short range (chip to chip), ultra-long range (inter satellite links), short range (Body area network), long range (inter building communication). Deep freshwaters. Low deployment time. Cost saving. Low operational cost. High fiber like bandwidth.	Atmospheric attenuation due to fog, smog, rain, smoke, etc.	Air-terrestrial-cities to satellite underground-underwater
MI	Transmits a signal across a shorter distance. It is lower power and lower cost, with no interference issues. Overcomes other features like scattering, multipath fading, frequency reuse, bandwidth limitation, and security issues. Applicable for shallow waters.	Only used for shorter distances. In VANET, V2V communication is RF, whereas inside vehicle to other device, best communication is MI.	Air-terrestrial-agricultural monitoring, mining, and transport tunnel monitoring Underground-leakage detection in water/gas/oil pipeline, detection of landslide, earthquake, volcano, tsunami, underwater-military submarine/surveillance, pollution monitoring underwater, scientific data collection
Molecular Communication	Information is encoded in molecules instead of EM signal. Drugs are designed with best molecular communication aiming to deliver drugs to their desired location at a controlled rate, minimizing the side effects. Solution of many diseases like cancer, genetic disorders, etc.	Complexity of human body and features like side effect, toxicology. Drug delivery system in human body.	Air-terrestrial transport of molecules through blood vessels in human body, cancer therapy Underground-underwater-body area network

underdeveloped countries, it is 25% to 30% per year. Implementation of Blockchain guarantees minimizes losses of foods while increasing profits for supply chain members [25]. It monitors the quality of food starting from agricultural land, to storage, to transportation. The monitored data throughout the process is visible to the customers. The RFID reader reads the food package with RFID and sensor. The package information is like a block which contains the RFID address, manufacturer, timestamp, sensor type, and sensor data [26]. RFID (radio frequency identification) is an automatic identification communication technology which identifies multiple, high-speed moving objects through a radio frequency signal. The block is interconnected with cryptographic function so that it creates an immutable digital database of food packets at each instance of time. Before Blockchain implementation we could see only the expiration date and manufacturer. Implementation of Blockchain in the food supply avoids food contamination, food waste, and losses due to spoilage [27]. Figure 4.2 shows how RFID is used in food supply chain management.

4.3.2 UNDERWATER SENSOR NETWORK SECURITY

Underwater sensor nodes help us in collecting data, monitoring oil, and minerals exploration, navigation, and surveillance applications. A huge number of sensors such as autonomous and unmanned (AUV, UUV) underwater vehicles are deployed to gather the scientific data [28]. The existing layers' security solutions for terrestrial use cannot be applied in an underwater WSN.

The major challenges are:

1. Underwater sensors are prone to failures because their topology changes frequently and are mobile due to flow of water. So underwater vehicles can move from one cluster to nearest cluster. Figure 4.3 explains this.

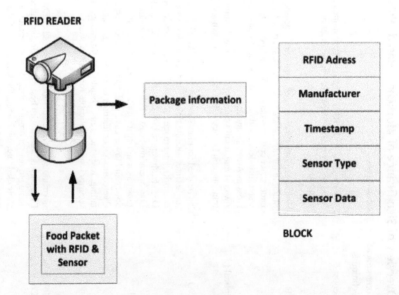

FIGURE 4.2 Food supply management using RFID.

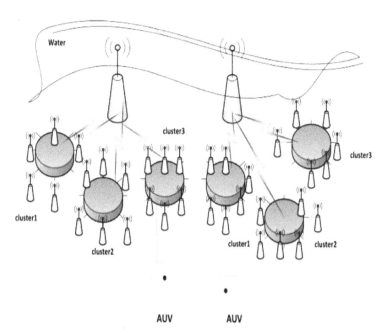

FIGURE 4.3 Movement of AUVs due to water flow.

2. Power is limited, and WSN devices cannot be recharged in water.
3. Radio frequency (RF) underwater is five orders of magnitude less than that in terrestrial.
4. Communication bandwidth is limited and affected by water temperature.

We propose an idea of implementing Blockchain technology in underwater sensor nodes such that confidentiality, authentication, integrity, availability, and denial of service attacks can be maintained. Figure 4.4 discuss the details of security issues and the Blockchain solution [29].

4.3.3 Telecom Roaming, Fraud, and Overage Management

Telecom services providers often come across issues related to roaming, fraud, and overage. Sometimes they don't have clear visibility of a subscriber's activity at roaming networks, which affects payment reconciliation [30]. Fraud subscribers can access the home service provider's network while cloning the roaming subscriber's identity. Blockchain solves all these issues. All the transactions are executed through a smart contract which improves quick payment and reduces fraud subscribers [31].

4.4 BLOCKCHAIN KEY CHARACTERISTICS

Pathan et al. analyzed the node identity, node compromising, location privacy, and key management threats. The other authors also discussed edge computing, better availability of services, securities, smart home, vehicle, cloud vs. fog computing,

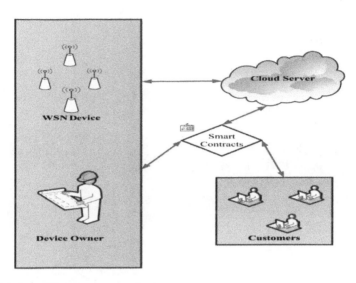

FIGURE 4.4   WSN device communication.

access management and identity management, data securities, authentication, and
fault tolerance; these are surveyed in [32].

4.5   NEED OF BLOCKCHAIN FOR DEVELOPING COUNTRIES

Blockchain is in its early stages, and it is a new technology which is attracting the
world's researchers, industries, merchants, and engineers. Thousands of implementation sectors have been proposed by engineers and researchers, and till today, there are
hundreds of small cryptographic coins.

4.6   REAL LIFE USES OF BLOCKCHAIN

4.6.1   Blockchain for Humanities Aid

In January 2017, the United Nations world food program started a project called
humanitarian aid. The project was developed in rural areas of the Sindh region of
Pakistan. By using Blockchain technology, beneficiaries received money and food,
and all types of transactions were registered on a Blockchain to ensure security and
transparency of this process. Swarm robotics with Blockchain implementation can
revolutionize many sectors starting from material delivery to precision farming.

4.6.2   Bitcoin Cryptocurrency

The most popular application of Blockchain, Bitcoin was launched in 2009 by an
unknown person called Satoshi Nakamoto [34]. Bitcoin is a peer-to-peer technology
which is not governed by any central authority or banks. Currently, issuing Bitcoins
and managing transactions are carried out collectively in the network. It is presently

the dominant cryptocurrency of the world. It is open source and designed for the general public, which means nobody owns control of the Bitcoin. In fact, only 21 million Bitcoin have been issued. Currently Bitcoin has a market cap of more than 12 billion dollars. Anyone can use Bitcoin without paying any process fees. If you are handling Bitcoin, the sender and receiver transact directly without using a third party.

4.6.3 INCENT CUSTOMER RETENTION

Incent is CRaaS (Consumer retention as a service) based on the Blockchain technology. It is a loyalty program which is based on generating tokens for business affiliated with its related network. In this system, Blockchain is exchanged instantaneously, and it can be stored in digital portfolios of the user's phone or accessed through a web browser. The smart City Dubai office introduced the Blockchain strategy. Using this technology, entrepreneurs and developers will be able to connect with investors and leading companies. The objective is to implement Blockchain base system which favors the development of various kinds of industries to make Dubai 'the happiest city in the world'.

4.7 DISADVANTAGES OF BLOCKCHAIN

Users may have fewer avenues to secretly address fraud, malpractice, or any kind of noncompliance of these networks. Security firms have also argued that it is possible for individuals and groups to insert malware into the Blockchain transaction. Data replication requires space, and the local copies of the Blockchain. Performances are therefore not yet comparable with databases. Immutability and transparency could harm a user's privacy and reputation. Every network node would store a copy of the Blockchain and could possibly access its content. Smart contracts cannot rely on external APIs. Every node should be able to process previous transactions and end with the same result as the other nodes. That is, information must be immutable. Consequently, data required by a smart contract should be first injected in the Blockchain. Oracles can enable this injection but require a strong reputation system or governance mechanism and need to be as robust as the Blockchain itself, not become the weakest part of the process. This probabilistic guarantee leads to security and performance issues, and attacks have been demonstrated.

4.8 CONCLUSION

The wireless sensor network (WSN) combined with Blockchain security will impact every aspect of our daily lives based on IoT, smart cities, smart vehicles, healthcare services, etc. However, the challenges still abound. The major challenges are low power and memory of wireless sensor devices. However, implementing Blockchain applications in WSN areas like body area network, IoT devices like smartphones, smart home appliance where humans are involved charging can be done at repeated interval of time or we can say power backup is not a constrained. In the last 50 years, the internet has boosted the world lifestyle. Now we say that Blockchain technology may improve the current lifestyle even more again. But how it will change is a general query that comes in mind. This chapter has included almost every basic knowledge for learning the Blockchain technology and it allows anyone to understand how Blockchain works, why it is useful, which fields require this technology, which effects make it worse, how it can fail, etc. Blockchain has shown its potential for transforming the traditional industry with its key characteristics: decentralization, persistency, anonymity, and auditability.

REFERENCES

1. K. Biswas and V. Muthukkumarasamy, *"Securing smart cities using Blockchain technology,"* in *2016 IEEE 18th International Conference on High Performance Computing and Communications; IEEE 14th International Conference on Smart City; IEEE 2nd International Conference on Data Science and Systems (HPCC/SmartCity/DSS).* IEEE, 2016, pp. 1392–1393.
2. K. Christidis and M. Devetsikiotis, "Blockchains and smart contracts for the internet of things," *IEEE Access*, vol. 4, pp. 2292–2303, 2016.
3. P. K. Sharma, S. Y. Moon, and J. H. Park, "Block-vn: A distributed Blockchain based vehicular network architecture in smart city." *JIPS*, vol. 13, no. 1, pp. 184–195, 2017.
4. A. Agah, K. Basu, and S. K. Das, "Security enforcement in wireless sensor networks: A framework based on non-cooperative games," *Pervasive and Mobile Computing*, vol. 2, no. 2, pp. 137–158, 2006.
5. A. Dorri, S. S. Kanhere, R. Jurdak, and P. Gauravaram, *"Blockchain for IoT security and privacy: The case study of a smart home,"* in *2017 IEEE International Conference on Pervasive Computing and Communications Workshops (PerCom workshops).* IEEE, 2017, pp. 618–623.
6. L. Fengi, H. Zhang, L. Lou, and Y. Chen, *"A Blockchain-based collocation storage architecture for data security process platform of WSN,"* in *2018 IEEE 22nd International Conference on Computer Supported Cooperative Work in Design (CSCWD).* IEEE, 2018, pp. 75–80.
7. M. S. Ali, M. Vecchio, M. Pincheira, K. Dolui, F. Antonelli, and M. H. Rehmani, "Applications of blockchains in the internet of things: A comprehensive survey," *IEEE Communications Surveys and Tutorials*, vol. 21, no. 2, pp. 1676–1717, 2018.
8. O. Novo, "Blockchain meets IoT: An architecture for scalable access management in IoT," *IEEE Internet of Things Journal*, vol. 5, no. 2, pp. 1184–1195, 2018.
9. H. Jang and J. Lee, "An empirical study on modeling and prediction of bitcoin prices with bayesian neural networks based on blockchain information," *IEEE Access*, vol. 6, pp. 5427–5437, 2018.
10. H. Gilbert and H. Handschuh, *"Security analysis of sha-256 and sisters,"* in *International Workshop on Selected Areas in Cryptography.* Springer, Ottawa, Canada, 2003, pp. 175–193.
11. E. Karaarslan and E. Adiguzel, "Blockchain based DNS and PKI solutions," *IEEE Communications Standards Magazine*, vol. 2, no. 3, pp. 52–57, 2018.
12. M. A. Khan and K. Salah, "IoT security: Review, blockchain solutions, and open challenges," *Future Generation Computer Systems*, vol. 82, pp. 395–411, 2018.
13. N. Kshetri, "Can blockchain strengthen the internet of things?" *IT Professional*, vol. 19, no. 4, pp. 68–72, 2017.
14. N. Kshetri and J. Voas, "Blockchain in developing countries," *IT Professional*, vol. 20, no. 2, pp. 11–14, 2018.

15. L. Lamport, R. Shostak, and M. Pease, "The byzantine generals problem," *ACM Transactions on Programming Languages and Systems (TOPLAS)*, vol. 4, no. 3, pp. 382–401, 1982.

16. X. Liu, "*A small java application for learning blockchain*," in *2018 IEEE 9th Annual Information Technology, Electronics and Mobile Communication Conference (IEMCON)*. IEEE, 2018, pp. 1271–1275.

17. M. Dabbagh, M. Sookhak, and N. S. Safa, "The evolution of blockchain: A bibliometric study," *IEEE Access*, vol. 7, pp. 19212–19221, 2019.

18. R. Yang, F. R. Yu, P. Si, Z. Yang, and Y. Zhang, "Integrated blockchain and edge computing systems: A survey, some research issues and challenges," *IEEE Communications Surveys & Tutorials*, vol. 21, no. 2, pp. 1508–1532, 2019.

19. T. D. Smith, "*The blockchain litmus test*," in *2017 IEEE International Conference on Big Data (Big Data)*. IEEE, 2017, pp. 2299–2308.

20. A. Kaushik, A. Choudhary, C. Ektare, D. Thomas, and S. Akram, "*Blockchain— literature survey*," in *2017 2nd IEEE International Conference on Recent Trends in Electronics, Information & Communication Technology (RTEICT)*. IEEE, 2017, pp. 2145–2148.

21. S. Yu, K. Lv, Z. Shao, Y. Guo, J. Zou, and B. Zhang, "*A high performance blockchain platform for intelligent devices*," in *2018 1st IEEE International Conference on Hot Information-Centric Networking (HotICN)*. IEEE, 2018, pp. 260–261.

22. G. Greenspan, "Smart contracts: The good, the bad and the lazy." [Online]. Available: Available: http://www.multichain.com/blog/2015/11/smart-contracts-good-bad-lazy/

23. EtherAPIs, "Decentralized, anonymous, trustless APIs, accessed on mar." [Online]. Available: https://etherapis.io/

24. A. Pandey and R. Tripathi, "A survey on wireless sensor networks security," *International Journal of Computer Applications*, vol. 3, no. 2, pp. 43–49, 2010.

25. F. Tian, "*An agri-food supply chain traceability system for China based on RFID and Blockchain technology*," in *2016 13th International Conference on Service Systems and Service Management (ICSSSM)*. IEEE, 2016, pp. 1–6.

26. S. Mondal, K. Wijewardena, S. Karuppuswami, N. Kriti, D. Kumar, and P. Chahal, "Blockchain inspired RFID based information architecture for food supply chain," *IEEE Internet of Things Journal*, vol. 6, no. 3, pp. 5803–5813, 2019.

27. U. Mukhopadhyay, A. Skjellum, O. Hambolu, J. Oakley, L. Yu, and R. Brooks, "*A brief survey of cryptocurrency systems*," in *2016 14th Annual Conference on Privacy, Security and Trust (PST)*. IEEE, 2016, pp. 745–752.

28. M. Conti, E. S. Kumar, C. Lal, and S. Ruj, "A survey on security and privacy issues of bitcoin," *IEEE Communications Surveys & Tutorials*, vol. 20, no. 4, pp. 3416–3452, 2018.

29. D. Mingxiao, M. Xiaofeng, Z. Zhe, W. Xiangwei, and C. Qijun, "*A review on consensus algorithm of blockchain*," in *2017 IEEE International Conference on Systems, Man, and Cybernetics (SMC)*. IEEE, 2017, pp. 2567–2572.

30. D. Puthal, S. P. Mohanty, P. Nanda, E. Kougianos, and G. Das, "*Proof-of-authentication for scalable blockchain in resource-constrained distributed systems*," in *2019 IEEE International Conference on Consumer Electronics (ICCE)*. IEEE, 2019, pp. 1–5.

31. D. Puthal, N. Malik, S. P. Mohanty, E. Kougianos, and G. Das, "Everything you wanted to know about the blockchain: Its promise, components, processes, and problems," *IEEE Consumer Electronics Magazine*, vol. 7, no. 4, pp. 6–14, 2018.

32. A.-S. K. Pathan, *Security of Self-Organizing Networks: MANET, WSN, WMN, VANET*. CRC Press, Boca Raton, FL, 2016.

33. W. She, Q. Liu, Z. Tian, J.-S. Chen, B. Wang, and W. Liu, "Blockchain trust model for malicious node detection in wireless sensor networks," *IEEE Access*, vol. 7, pp. 38947–38956, 2019.

34. A. K. Sharma, S. Yadav, S. N. Dandu, V. Kumar, J. Sengupta, S. B. Dhok, and S. Kumar, "Magnetic induction-based non-conventional media communications: A review," *IEEE Sensors Journal*, vol. 17, no. 4, pp. 926–940, 2016.

5 Modified Artificial Fish Swarm Optimization Based Clustering in Wireless Sensor Network

Biswa Mohan Sahoo
Amity University, Uttar Pradesh, India

Ramesh Chandra Sahoo
Utkal University, Bhubaneswar, Odisha, India

Nabanita Paul
International Institute of Information Technology, Bangalore, India

Abhinav Tomar
Netaji Subhas University of Technology, Delhi, India

Ranjeet Kumar Rout
National Institute of Technology, Srinagar, Jammu and Kashmir, India

CONTENTS

DOI: 10.1201/9781003145028-5

5.1   INTRODUCTION

In computing and communication technology, ambient intelligence is the upcoming trend. Battery powered nano sensors, wireless networks, and the main elements of this issue are the merged intelligent software. The wireless sensor network is the developing approach which, based on several devices in terms of sensor nodes, can be intently employed in any ordinary and obscure environments, to monitor with high accuracy physical phenomena or events such as ecological monitoring, natural surroundings monitoring, battlefield surveillance, and organic attack detection. The event is reported whenever a sensor node feels any earthly movement (pressure, heat, sound, areas having some magnetic properties, vibration, etc.), to one of these sites which gathers and transfers the message and then provides a certain response. Each one of these devices is called a sensor node. Typically, WSN is an infrastructure-less wireless network that may be composed of hundreds or thousands of low-cost sensor nodes distributed in a geographical area either with a fixed location or deployed randomly for monitoring purposes. A special node called a base station (also referred to as a sink) acts as a bridge between a WSN and remote users. The primary task of a base station is to gather data from other sensor nodes and disseminate sensed data to another network for further processing. The sensor nodes near to the base station are quickly depleted of energy, as they are transmitting data to nodes that are far away from the sink node. A WSN is framed by tiny and low-cost sensors that capture and maintain track of environmental data. These sensor nodes are employed in an area where some occurrence took place. There are several challenges in WSN; for instance, power management, distance management, real-time challenges, design issues, topological issues, and so on. Among them one of the vital challenges in WSN is the optimization of consumption of energy. According to the research [1–5], the usage and the optimization of energy consumption by the sensor in the cluster-based networking model can be done more effectively. Furthermore, this algorithm integrates the decomposed cluster into the target partition. The only key goal of this algorithm is to detect the data sets or cluster into similar groups and to minimize the objectives in parallel. By this process, sometimes the number of cluster becomes greater than the processor which increments the scheduling length and the longevity of the cluster.

Data transmission is an indispensable responsibility for the wireless sensor network. The main goal of WSN is to segregate the essential information of source nodes embedded in the data and deliver it to the destination, for data transmission clustering is one of the simpler and well-known mechanisms. In simple language, for a wireless sensor network, a cluster is a set of nodes. Each cluster communicates with each other and works towards the targeted goal. Nodes from each cluster can be added and removed from the cluster anytime and among the nodes one of them is employed as a leader, named cluster head (CH). However, the WSN faces an abundance of challenges at the moment of clustering. As stated before, nodes can be added dynamically, so sometimes it may happen that the number of cluster becomes greater than the processor. Therefore, several complications, like more energy consumption, delay in data transmission, etc., have been raised. In fact as the number of cluster becomes higher during the communication, the collected data set also becomes huge, which implies the graph of redundancy in data that arises. It is too difficult to identify the valuable and error-free information from a large set of data, because the collected data through the source node in cluster is large in general, wherein a data set contains a plethora of information that contains redundant and irrelevant features, which causes the delay in execution and consume too much energy on the network [1,6,7].

The date gathering approach for energy balancing in the wireless sensor network encompasses some of the following benefits:

a. Compressive data aggregation methods may lead to recovery the process of data.
b. As cluster head cumulates the collected data and transmits only meaningful data to the sink, the energy of the nodes can be saved.
c. CHs preserve the local route circumstance for the other CHs.

The rest of the manuscript is organized as follows. In Section 5.2, related work to the proposed technique is discussed. In Section 5.3, the proposed technique has been discussed that describes the whole process of MAFS and K-mean with discussion to its clustering. In Sections 5.4 and 5.5, performance metric the results and simulation analysis are discussed comprehensively. Finally, the conclusion and future scope is reported in Section 5.6.

5.2 RELATED WORK

In this literature work, the different methods for data gathering and cluster head selection employed in the fish swarm algorithm are discussed. Heinzelman et al. [8] proposed a popular clustering method named Low-Energy Adaptive Clustering Hierarchy (LEACH) in their study (Heinzelman, Chandrakasan, & Balakrishnan, 2000). The algorithm works with dual phases: the first is the set-up phase, and another is the steady phase. In this algorithm every node gets a chance to be a CH by satisfying the threshold formula, i.e. if the number of node (n) is less than threshold value T(n), that node can be the CH of the cluster. In the LEACH algorithm, the CH aggregates data and directly or indirectly sends it to the base station (BS), implying it does not need a central control. There are several advantages. For instance, as nodes send data to the CH through single-hop communication and whole collected data are segregated merely by CH which straight shows the reduction in energy consumption [9]. However, as we said before, the selection of CH is too random to keep a record regarding the number of the employed CH, and also the division of the cluster is random; that is why the position and the number of nodes in the cluster is uneven which causes the greater energy consumption. In fact, as the CH sends data directly or indirectly to the BS, failure of any CH makes that cluster useless.

Several researchers have used the AFSA method for deploying and tracking of WSN and found a significant performance when compared with other algorithms. Basically, an optimization method is applied to a problem to either maximize or minimize the output by adjusting its features to select an optimal solution out of many possible solutions. In this context, swarm intelligent algorithms are worked iteratively on a population by which they can improve their individual positions after every iteration and subsequently reach a better position to achieve the goal [10]. In [11], Yiyue et al. proposed an optimized AFSA (OAFSA) to increase network coverage by maximizing the objective function for global optimization. In the start of the algorithm, sensor nodes are moved virtually, and after getting the best location at the end of the algorithm, the nodes are moved physically to that location. Ma and Xu, [12], presented a Niching particle swarm optimization algorithm with dual cluster heads for energy awareness distance-based clustering by considering residual energy and distance metric. W. Zhao et al., [13], proposed a fish swarm algorithm for underwater WSN to optimize the mobile node formation and gathering, preying and chasing behaviors of fish swarm are used to achieve maximum search area by controlling their movements. Huang and Chen, [14], proposed an improved AFSA algorithm with a hybrid behavior selection method. The authors used swallowed behavior for speeding-up convergence and breeding behavior for improving global optimization.

In [15], Sengottuvelan and Prasath proposed an improved AFSA method (BAFSA: breeding artificial fish swarm algorithm) for CH selection in WSN with improved network lifetime as compared to LEACH and GA. In this research, solutions are considered as binary values and the Hamming distance metric is used to measure distance. Salawudeen et al., [16], proposed a modified AFSA named weighted artificial fish swarm algorithm (wAFSA) for network coverage and nodes mobility. To get the better objective function, the authors introduced an inertial weight in the original AFSA algorithm that can pick its parameters before deployment of WSN. The performance shown in this study outperforms over AFSA that enables search in a larger space before it converges due to the adaptive nature of the inertial weight.

However, deployment of mobile nodes and utilizing their coverage area effectively is a challenging task for researchers in WSN.

5.3   PROPOSED METHODOLOGY

The major objective in this chapter is to develop a data gathering technique in an efficient manner with optimal path selection. Here, the energy consumed cluster head selection process is carried out with the aid of weighted k-means clustering (WKMC) and the modified artificial fish swarm (MAFS) algorithm. The proposed methodology comprises for two phases, described as follows:

- Weighted k-means clustering algorithm to group the sensor node.
- Modified artificial fish swarm algorithm-based cluster centroid selection.

5.3.1   CLUSTERING THE SENSOR NODE

Clustering in WSN is the process of accommodating sensor nodes in small groups based on some mechanism with its cost and memory, which plays an important role in handling many problems like efficient energy consumption, scalability, and lifetime of sensor network. The clustering operation comprises three phases: cluster head (CH) selection, cluster formation, and transmission. To demonstrate clustering in a WSN, let us say $SN = \{sn_1, sn_2, ..., sn_i\}$ are sensor nodes deployed in a WSN, where sn_i is the ith sensor node and its position can be represented by its x and y coordinates of a 2D plane of the spatial area. The grouping of these sensor nodes can be done by the following steps.

5.3.1.1 Weighted k-Means Clustering Algorithm

The popular k-means clustering algorithm is an unsupervised learning method that partitions the data set into k distinct subgroups of non-overlapping sensor nodes. In this work a variant of k-means named the weighted k-means clustering method is analyzed to divide a given set of data points into different clusters. In this way the size and difficulty of each subset of data points can be minimized, and the efficiency and effectiveness of this model can be improved. In this method data points are associated with a set of weights, and the Euclidean distance metric of centroid is measured along with weights of data points. Thereby the weighted k-means algorithm helps to shrink effects of immaterial attributes and impersonate the semantic information of objects. The working procedure of weighted k-means clustering is as follows.

Weighted k-means clustering algorithm
Input: N nodes data set and number of clusters (k)
Output: Centroid of k clusters

Step 1: Initialize k clusters.

Step 2: Select the centroid of 1,2,..., k from the data set randomly.

Step 3: Consider their weight values (w).

Step 4: Find the Euclidean distance between the centroid as:

$$ED_{ij} = w \times \left(s_i - c_k \right)^2 \tag{5.1}$$

Step-5: Update the centroid.

Step 6: Terminate the procedure if newly obtained centroid is nearer to the old centroid; else, go to step 4.

From the weighted k-means clustering process, the number of clusters obtained are utilized for further processing phases.

5.3.2 CLUSTER HEAD SELECTION

Each and every node in the network comes under one of the cluster groups. The centroid of clusters is selected based on storage capacity, distance, energy, etc. The node with maximum capacity, minimum distance, and low energy is to be chosen as the centroid of that cluster. Data are gathered from all the nodes via multi-hop or directly to the cluster head. The collected information is transferred to the mobile sink. Here, in this research, the cluster centroid is being selected optimally with the modified fish swarm algorithm.

5.3.2.1 Modified Artificial Fish Swarm Algorithm (MAFS)

The modified artificial fish swarm algorithm (MAFS) is one of the well-known swarm intelligence optimization algorithms that work in light of populace and stochastic search. MAFS is an arbitrary searching optimization algorithm propelled by a fish's practices, for example, searching for food, swarming, and tailing others. It is great at staying away from the neighborhood ideal and searching for the global optimum inferable from its versatile limit in the parallel search of arrangement space through reproducing these practices in nature.

The working for MAFS is presented in Figure 5.2 which shows the flow of the algorithm. Here, the cluster center is opted optimally by using various features. In light of the behavior depiction of the previously mentioned artificial fish, each artificial fish searches its suitable conditions and its acquaintances for selecting a proper behavior to move at the speediest towards the optimal path. Finally, the artificial fish bring together few neighborhood extrema. The practices of artificial fish incorporate prey behavior and swarm behavior, along with follow behavior. These practices are depicted in the underneath area. The following are the steps of MAFS.

5.3.2.1.1 Initialization

Input parameters like weight α and β are to be initialized and can be expressed as α_i and β_i as an initial solution of artificial fish, where i represents the number of solutions. Here, N be the total number of nodes in the cluster.

5.3.2.1.2 Oppositional Behavior

From the opposition based learning (OBL), the initial and the dependable reverse solution is created. This will increase the effectiveness of our proposed algorithm to convalesce the accuracy of solutions.

5.3.2.1.2.1 Prey Behavior

This is a vital biological method that tends to the food. The fish update the position based on the more food-concentrated region.

$$Ni + 1 = Ni + \text{rand}(r1)\frac{Nj - Ni}{\|Nj - Ni\|}, \text{fitness} \, j < \text{fitness} \, I \tag{5.2}$$

$$N_{i+1} = N_i + \text{rand}(r2) \tag{5.3}$$

where
- N_i describes the position of fish
- N_j describes the new position update

5.3.2.1.2.2 Swarm Behavior

All fish will normally swim based on the food concentration and crown in order to avoid danger:

$$Ni + 1 = Ni + \text{rand}(r1)\frac{Nc - Nt}{\|Nc - Nt\|}, \text{fitness} \, c < \text{fitness} \, t \, \text{and} \left(\frac{\eta_s}{\eta}\right) < \delta \tag{5.4}$$

$$Nc = \frac{1}{n} \sum_{i=1}^{n} Nt, j \qquad (5.5)$$

5.3.2.1.2.3 Follow Behavior

If the fish finds more food in another location, and that location is not crowded, that means the fish will immediately swarm to that position.

$$Ni+1 = Ni + rand(r1)\frac{Nbest - Nt}{\|Nbest - Nt\|}, \text{ fitness best} < \text{fitness } t \text{ and} \left(\frac{\eta_s}{\eta}\right) < \delta \qquad (5.6)$$

where, fitness $_{best}$ denotes as best fitness.

5.3.2.1.2.4 Termination Criteria

The process continues until the best fitness value is obtained. The node with best fitness is chosen as the optimal cluster centroid.

The pseudo code for artificial fish swarm is illustrated as follows:

Pseudo code for modified artificial fish swarm
    For each artificial fish I, where I ε [1,2,.......n]do
    Initialize N_i
    Reserve the initial solution using opposition-based learning
End for
Repeat
//searching behavior
For each artificial fish i do
    Obtain N_{i+1} with equation
End for
//swarming behavior
Obtain center position N_c with equation
For each artificial fish i do
    If $f(N_c) \geq f(N_{i,j})$, then
        If $N_{i,j}$ is the best fish in swarm i then
$N_{i,j} = N_c$
Else
Applying Equation (5.4) on $N_{i,j}$
    End if
  End if
 End for
//chasing behavior
For each artificial fish i do
    Obtain best fish N_{best} with the Equation (5.6)
End for

To optimize energy consumption, the unused cluster centroids are removed, and cluster heads are used to transmit data based on optimal weights.

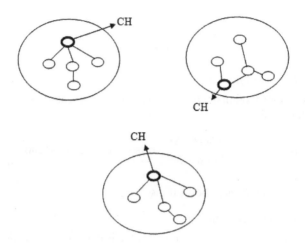

FIGURE 5.1   Cluster head selection.

5.4   PERFORMANCE METRICS

Following performance measures are used to analyze the efficiency of the proposed method.

5.4.1   END-TO-END DELAY

The end-to-end delay of a network can be defined as the time taken to transmit a bit to the destination and the unit used to measure this parameter is in seconds.

5.4.2   THROUGHPUT

Throughput can be defined as the amount of data transmitted from a source node to destination in one second and the unit to measure this parameter is kb/s.

$$Throughput = \frac{Total\ data\ transmitted\ in\ kb(s)}{Transmission\ time\ in\ second(s)}$$

5.4.3   NETWORK LIFETIME

Lifetime of WSN can be defined as the initial node of the network transmit a packet based on duration or time utilization may reduce from the energy limit.

5.5   COMPARATIVE ANALYSIS

The effectiveness of the proposed data gathering method is analyzed with existing algorithms so as to test whether the recommended work is the better one. For evaluation purposes, we are now analyzing current procedure as BAFSA and WAFSA methods [9,15] and data gathering without optimization. The detail description is exemplified in the following segment.

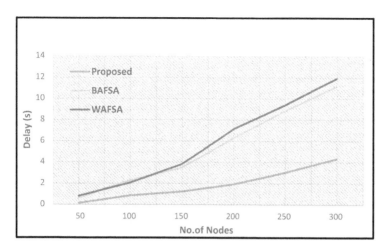

FIGURE 5.2   Comparison between delay and nodes.

Figure 5.2 demonstrates the comparison result of the proposed method with BAFSA and WAFSA, and it has been observed that the delay value is increased when the number of nodes increases. For 50 numbers of nodes, the delay value of the proposed method is a minimum value when compared to the BAFSA and WAFSA methods. The delay value of this proposed method is observed to be 0.1034s. Likewise, we are calculating the delay value for 100, 150, 200, 250, and 300 nodes where we found minimum delay of this proposed method as compared to the anticipated method for all scenarios of nodes.

Figure 5.3 represents the throughput comparison graph of our proposed method with BAFSA and WAFSA methods. It has been observed that the throughput of our proposed method is recorded as 1914, 2100, 2267, 3178, 4219, and 5145 Kbps for 50,

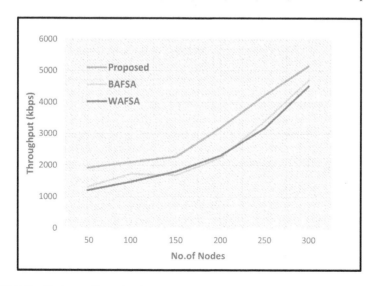

FIGURE 5.3   Nodes vs. throughput.

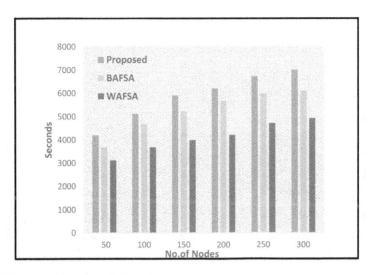

FIGURE 5.4   Nodes vs. network lifetime.

100, 150, 200, 250, and 300 nodes respectively. When compared with the BAFSA and WAFSA protocols, the proposed method attains the maximum throughput value for both 50 and 100 nodes. Correspondingly, we are calculating the throughput value for node 150, 200, 250, and 300 nodes. It has been seen that, in our proposed method, we achieved maximum throughput as compared with the anticipated method for all nodes.

Figure 5.4 represents the comparison graph for network lifetime between the proposed method with BAFSA and WAFSA. For 50, 100, 150, 200, 250, and 300 nodes, the proposed network period values are 4200, 5123, 5898, 6213, 6734, and 7008s respectively, which is maximum value when compared with the existing methods. From our experimental results, it has been observed that the proposed method utilizes maximum of network lifetime as compared to the existing methods.

Figure 5.5 demonstrates the energy consumption comparison of our proposed method with offered and surviving systems. Energy application of the forecast

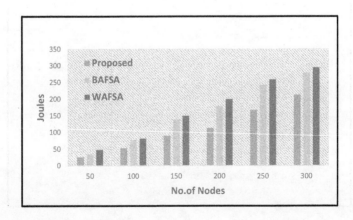

FIGURE 5.5   Nodes vs. energy.

TABLE 5.1
Efficiency Analysis of Our Proposed Methodology

No. of Nodes	Delay (s)	Energy Consumption (Joule)	Network Lifetime (s)	Throughput (Kbps)
50	0.1034	25	4200	1914
100	0.80269	52	5123	2100
150	1.1865	89	5898	2267
200	1.8828	112	6213	3178
250	2.9762	167	6734	4219
300	4.2712	212	7008	5145

progression is under the outdated approach. As a cluster centroid is responsible for data communication, transmitting whole data to mobile sink from cluster head in our proposed process needs minimum energy as compared to conventional approaches. This can be clearly seen from our achieved experimental results of the proposed technique that attains the minimum energy value when compared to the conservative methods for all node scenarios.

5.5.1 PERFORMANCE EVALUATION

Table 5.1 shows the assessment of the proposed system with different performance metrics such as delay, energy consumption, network lifetime, and throughput when running the network with 50, 100, 150, 200, 250, and 300 nodes.

5.6 CONCLUSION

This chapter presented an effective data gathering technique and enhanced the optimization of WSN by means of a modified artificial fish swarm algorithm (MAFS). The system was imitation in MATLAB® R2018a environment, and results obtained from the proposed modified method show the superiority over the standard BAFSA and WAGSA methods. The superiority of the modified algorithm is attributed to the adaptive and dynamic behavior of the modified algorithm in comparison the original BAFSA and WAGSA. However, there is always an opportunity to enhance the proposed work further by extending the various clustering algorithms to aggregate mobile sink nodes effectively.

REFERENCES

1. Akila, I. S. and Venkatesan, R., "A fuzzy based energy-aware clustering architecture for cooperative communication in WSN," *The Computer Journal*, Vol. 59, No. 10, pp. 1551–1562, 2016.
2. Madhumathy, P. and Sivakumar, D., "Enabling energy efficient sensory data collection using multiple mobile sink," *China Communications*, Vol. 11, No. 10, pp. 29–37, 2014.
3. Sahoo, B. M., Amgoth, T., and Pandey, H. M., "Particle swarm optimization based energy efficient clustering and sink mobility in heterogeneous wireless sensor network." *Ad Hoc Networks*, Vol. 106, p. 102237, 2020.

4. Sahoo, B. M., Rout, R. K., Umer, S., and Pandey, H. M., "*ANT colony optimization based optimal path selection and data gathering in WSN.*" In *2020 International Conference on Computation, Automation and Knowledge Management (ICCAKM)*, pp. 113–119. IEEE, Dubai, 2020.

5. Sahoo, B. M., Pandey, H. M., and Amgoth, T. "GAPSO-H: A hybrid approach towards optimizing the cluster based routing in wireless sensor network." *Swarm and Evolutionary Computation*, 60, 100772, 2021.

6. Sahoo, B. M., Gupta, A. D., Yadav, S. A., and Gupta, S., "*ESRA: Enhanced stable routing algorithm for heterogeneous wireless sensor networks.*" In *2019 International Conference on Automation, Computational and Technology Management (ICACTM)*, pp. 148–152. IEEE, London, UK, 2019.

7. Qiao, J. and Zhang, X., "Compressive Data Gathering Based on Even Clustering for Wireless Sensor Networks," *IEEE Access*, Vol. 6, pp. 24391–24410, 2018.

8. Heinzelman, W. R., Chandrakasan, A., and Balakrishnan, H., "*Energy-efficient communication protocol for wireless microsensor networks.*" In *Proceedings of the 33rd Annual Hawaii International Conference on System Sciences*, p. 10, IEEE, Maui, HI, USA, 2000.

9. Sahoo, B. M., Pandey, H. M., and Amgoth, T., "*A Whale Optimization (WOA): Meta-Heuristic based energy improvement Clustering in Wireless Sensor Networks.*"In *2021 11th International Conference on Cloud Computing, Data Science & Engineering (Confluence)*,pp.649–654,Noida,India,2021.doi:10.1109/Confluence51648.2021.9377181

10. Azizi, R., Sedghi, H., Shoja, H., and Sepas-Moghaddam, A., A novel energy aware node clustering algorithm for wireless sensor networks using a modified artificial fish swarm algorithm, (2015). arXiv preprint arXiv: 1506.00099.

11. Yiyue, W., Hongmei, L., and Hengyang, H., "*Wireless sensor network deployment using an optimized artificial fish swarm algorithm.*" In *2012 International Conference on Computer Science and Electronics Engineering (ICCSEE)*, pp. 90–94, Hangzhou, China, 2012.

12. Ma, D. and Xu, P., "An energy distance aware clustering protocol with dual cluster heads using niching particle swarm optimization for wireless sensor networks." *Journal of Control Science and Engineering*, Vol. 2015, Article ID 849281, 2015.

13. Zhao, W., Tang, Z., Yang, Y., Wang, L., and Lan, S., "Cooperative search and rescue with artificial fishes based on fish-swarm algorithm for underwater wireless sensor networks." *The Scientific World Journal*, Vol. 2014, 2014.

14. Huang, Z. and Chen, Y., (2013). "An improved artificial fish swarm algorithm based on hybrid behaviour selection." *International Journal of Control and Automation*, Vol. 6, No. 5, pp. 103–116.

15. Sengottuvelan, P. and Prasath, N., "BAFSA: Breeding artificial fish swarm algorithm for optimal cluster head selection in wireless sensor networks." *Wireless Personal Communications*, Vol. 94, No. 4, 1979–1991, (2017).

16 Salawudeen, A. T., Abdulrahman, A. O., Sadiq, B. O., and Mukhtar, Z. A., "*An optimized wireless sensor network deployment using weighted Artificial Fish Swarm (wAFSA) optimization algorithm.*" In *International Conference on Information and Communication Technology and its Applications*, pp. 203–207, Minna, Nigeria, 2016.

6 Survey

Data Prediction Model in Wireless Sensor Networks Using Machine Learning and Optimization Methods

S. Ramalingam and K. Baskaran

Alagappa Chettiar Government College of Engineering and Technology, Karaikudi, Tamil Nadu, India

Umashankar Subramaniam

Prince Sultan University, Riyadh, Saudi Arabia

CONTENTS

6.1   INTRODUCTION

A WSN collects many small sensor nodes applied in different areas such as healthcare systems, energy monitoring, indoor and outdoor environmental surveillance, object monitoring, crop monitoring, robotic exploration, military applications, etc. Sensors are distributed in space with the freedom to govern themselves or control their affairs [1]. These sensors detect the environmental conditions and transmit the detected data to the receiving sensor node (SN) over the WSN network. The receiving node collects all data sets for further processing. This sensor node consumes more power for communication, data processing, and detection. The inclusion of communication causes high energy consumption.

Since the wireless sensor node is kept in a difficult-to-reach position, it is inconvenient to change the battery regularly. This high power consumption problem can be reduced by reducing data sent to the destination node. Various data reduction techniques have been developed to overcome these problems.

Algorithmic approaches, the stochastic process, and time series prediction are the three subclasses of data prediction techniques [2]. Forecasting data with minimal RMSE is a technique to minimize high power consumption without compromising data quality [3]. The data aggregation method is performed using the internet technique in the path where the data is transferred to the receiving node to transform a large amount of collected data into less detailed refined data [4–6]. Data compression, data prediction, and internet processing are the techniques developed to decrease the number of data transmissions. The data compression technique is applied to limit the maximum number of data received in sink node [7]. The data compression technique is used when the last measurement is not required for the WSN application [8]. In this chapter, we review and analyze the different data prediction algorithms for the wireless sensor network. WSNs suffered mainly from energy sources; several developed routing algorithms to minimize energy consumption. In a WSN, routing is defined as data transmission from the source-receiver node or the receiver-sensor node. To design a wireless sensor network, the designer must address challenges such as location, implementation, planning, security, data aggregation, and QoS instead of routing.

Each sensor node adds a large volume of data to be sent, processed, and received in WSN. Due to sensor bandwidth and power constraints, processing and decision-making are difficult [9]. To more efficiently manage data, machine learning is used. The WSN data prediction schemes are shown in Figure 6.1. We are unable to correct the data after collecting it. Several sectors use machine learning in these situations, including industry, medicine, and the military. To discover valuable data, they apply machine learning.

6.2 MACHINE LEARNING (ML) ALGORITHMS

The ML algorithm is a part of artificial intelligence that enables machines to learn from training data. The algorithm employs computational methods to directly analyze data sets while relying on standard equations as a model. As the volume of training data available grows, the methods are constantly updated to enhance performance.

ML algorithms are significant tools that can solve large optimization issues with sensor data or integrated inputs independently. It encourages effective decision-making and successful execution with limited external interference in a real-world implementation. Technologies for machine learning are continuously emerging and are widely applied in almost all fields. Similarly, their implementations have unique limitations [10]. There are mainly two different types of ML algorithms available: supervised learning and unsupervised learning, as shown in Figure 6.2.

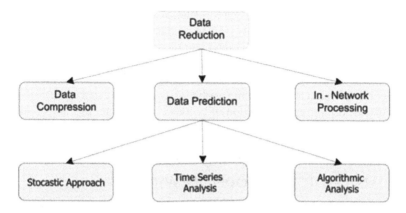

FIGURE 6.1 Data-driven schemes in WSN to reduce the energy consumption.

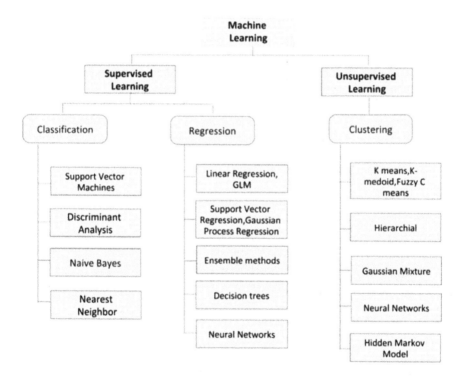

FIGURE 6.2 Machine learning algorithms.

The ML algorithm uses a collected data set used to train the model to predict the desired accurate data. Choosing an algorithm depends on the amount, type, and understanding of the data. Supervised learning methods are also used in the intellectual study of IoT data in various fields [11–13]. Regression is monitored by ML engineers who forecast ongoing responses such as stock prices, energy demand fluctuations, and time series sensors.

Four techniques have been selected for detailed discussion as they are relevant to applying crop yield predictions [14]. SVM prediction models evaluate a sequence that deviates with a value not wider than a limited percentage of model parameters from the calculated data that minimizes error susceptibility. It is acceptable for large data where many predictive variables are present [15,16]. The Ensemble Learning (EL) algorithm is used to improve the prediction performance and enhance the accuracy of data. To produce a collection of accurate and distinct hypotheses, the first objective is to independently find each hypothesis. The algorithms for bootstrap and bagging produce a good fit for the data [17].

6.3   DATA PREDICTION MODELS IN WSN

6.3.1   PCA

Murad A. Rassam et al. present the suggested high-quality data reduction model regarding principal component analysis (PCA). WSN is also used to monitor physical and environmental factors like humidity, temperature, vibration, voltage, and light. Fortunately, when transferring such data to the sink, the large size of the recorded data, particularly for multivariate application areas, increases energy usage and decreases the sensor's lifespan. A reliable data reduction technique is therefore needed to minimize energy usage during transmission of data.

To minimize the coordinates represented by the data and increase the new coordinates' variance value, the principle that motivates PCA data reduction is used. The algorithm consists of two phases, the preliminary and the reduction phases. In terms of communication and computational complexity, the approach's efficiency was calculated, while reliability has been evaluated based on the accuracy of the reconstruction of the original data [18].

At present, PCA has been used to process large yet correlated sensor data. However, it is not easy to adapt the conventional PCA approach to the WSN's complex conditions. The goal of the PCA technique is to reduce the use of redundant data transmission network resources. Meanwhile, we ensure the accuracy of the information received at the sink. Data redundancy problems and data error can be managed by a new R-PCA-based algorithm [19,20].

6.3.2   ARIMA Prediction Model

ARIMA forecasting models are based on statistical principles and calculations. A wide range of time series behaviors is modeled with ARIMA modeling. AR, MA, and ARMA are the three basic models of ARIMA. The simultaneous application of normal differentiation for AR and MA is termed ARIMA. Here, the integrated and referenced differentiation method. Intervention data, univariate equivalent time series data, and functional data are analyzed and predicted using the ARIMA prediction model. The results obtained in this work show that by applying different prediction algorithms to data sets such as air temperature, humidity, and soil temperature, RMSE is calculated concerning ARIMA lower than other algorithms [21,22].

6.3.3 MULTIPLE LINEAR REGRESSION MODELS

Multivariate temporal and spatial correlation approaches are used in [23] to enhance predictive data precision and data reduction for WSN. Prediction data-based data transfer is used in the WSN to save energy by reducing traffic. Linear regression and MLR were performed to evaluate the proposed method's prediction quality. The prediction accuracy performance is evaluated using the remaining sum of squares (SSerr) and determination coefficient (R^2). Compared to MLR, the prediction accuracy was low in linear regression.

6.3.4 SUPPORT VECTOR MACHINE

This work presents a support vector machine (SVM) application for forecasting weather data. The time-series data prediction model was proposed, with different location maximum temperature analyzed using SVM. Support vector regression is distinct from traditional methods of regression. For various orders, the output of SVM was compared to that of MLP [24,25]. The principle of determining a hyperplane that divides a data set into two groups is based on SVMs. The accuracy level of SVM is high compared to other machine learning algorithms. We used the SVM algorithm to classify meteorological parameters to predict precipitation and apply it to binary classification. It supports the vector machine algorithm to predict whether it will rain or not [26].

A new technique focuses on PV intensity data and partial weather data for forecasting weather forms. With this approach, instead of meteorological instruments, weather categories are extracted from data analysis. Better error detection is achieved using SVM and comparing the expected and actual times. For training multi-class predictors, a direct SVM sets up the model for weather forecasting. We used limited weather data when predicting to increase precision. At a much lower expense, this meteorological data can be collected. The SVM models we have developed can assist with error detection, but we cannot decide the outcome of the solar cell system's condition, or the fault position, or the type of fault. To make a further decision [27], information on other aspects is required.

6.3.5 ENSEMBLE METHODS

Predictive power for the short-term solar radiation forecast has been proposed for very good ensemble techniques like boosted trees, random forest, bagged trees, and generalized random forest. Ensemble learning is a very effective tool that can increase a weak model's effectiveness in making accurate predictions. Four models of ensemble learning have been developed for hourly global solar radiation using meteorological factors. Compared to other machine learning algorithms, ensemble learning models provide superior predictive performance [28]. The prediction of solar energy production has been proposed based on photovoltaic meteorological data using random forest algorithms.

A two-step modeling method that compares sudden meteorological variables with confirmed weather predictions is proposed in this research. Empirical results indicate

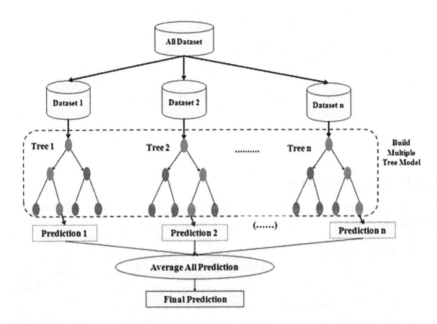

FIGURE 6.3   Ensemble learning algorithm.

that this technique, irrespective of the machine learning algorithms used, reinforces a simple process with high margins [29]. In Figure 6.3, the ensemble learning model is shown.

6.3.6   Artificial Neural Network (ANN)

In several energy systems, predicting solar radiation plays an essential role. This chapter aims to implement an affordable solar radiation measurement system using artificial neural networks (ANN). Using the post-propagation algorithm, solar radiation logs were used to train the ANN (online) parameters. Driven NN has been used to forecast solar radiation by calculating the photovoltaic module's open-circuit voltage and short circuit current. The model based on the photovoltaic module's actual behavior is more accurate than any other mathematical method for solar radiation calculation.

The proposed approach prevents using a costly solar radiation measurement pyranometer, offering an accurate and cost-effective adaptive approach for measuring solar radiation. Due to fault conditions, the featured photovoltaic monitoring device will decide whether the photovoltaic module has deteriorated. For small artificial neural networks, the monitoring system implements an effective reference model. This reference model for artificial intelligence is used to predict the standard photovoltaic module output power operating under various environmental conditions [30,31].

6.3.7   Multilayer Perceptron (MLP)

We propose a multilayer perceptron (MLP)-based method to predict a building's energy consumption from the data obtained from WSN [32]. The vector of characteristics of

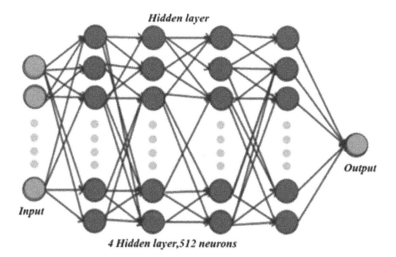

FIGURE 6.4   MLP with 4 hidden layer.

the input is mapped to a single output reflecting the house's energy consumption via a sequence of hidden levels. Based on the original level's outcome, each network hidden layer provides a specific input representation. The network will establish a strong representation of high input with such a degree of subtraction. Energy forecasts, therefore, play an important part in energy reduction. Figure 6.4 demonstrates the modeling of a multilayer perceptron neural network and energy prediction system.

6.3.8   Long Short Term Memory (LSTM)

The suggested LSTM uses as input variables of the hourly weather forecast for the day. The values of hourly radiation evaluated for the day occurring on the same scheduled day. Therefore, the prediction problem is considered a problem of standardized outcomes with multiple outcomes being expected simultaneously. In Figure 6.5, the standardized prediction performance model is based on LSTM networks. In terms of root mean square error (RMSE), the proposed algorithm is 18.34% more precise than BPNN [33].

A new method for predicting multi-function multi-node (MNMF) data is based on long-term two-way array memory (LSTM). In the quadratic approach, the data quality is improved. Therefore, to extract and learn abstract properties from sensory data, the two-way LSTM network is used. Finally, conceptual technologies can be used for predicting data by supporting the neural network linking layer. The suggested MNMF approach simulation results are more effective than existing statistical error parameter algorithms [34].

6.4   HYBRID MODELS

6.4.1   PSO-SVM

In this chapter, the PSO-SVM model was proposed for weather data prediction for WSN using various parameters to predict dependent variables such as temperature,

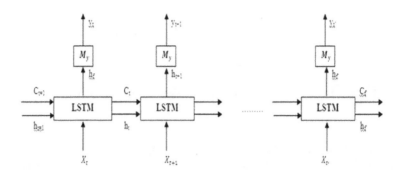

FIGURE 6.5 LSTM model.

wind speed, humidity, and wind direction, in different climate change environments. A hybrid particle optimization with a support vector machine (PSO-SVM) is introduced to enhance data prediction accuracy. The parameter selection has a vital effect on the accuracy of the SVM's prediction, and it is recommended that the PSO identify the optimal parameters for the SVM. To train a model, the suggested model was used: historical data to forecast rainfall. Combining the PSO algorithm and the SVM model can be solved effectively, and the optimum SVM parameter can be accurately found.

This hybrid algorithm has proven to be an efficient method for determining precipitation forecasts. The SVM is a type of algorithm for ML with a large degree of nonlinear issues. This hybrid model was created in conjunction with traditional network optimization, the ant colony algorithm, and the genetic algorithm. Experimental outcomes show that the algorithm for PSO is more precise and efficient. This is ideal for multifactorial analyses and is specifically important for existing problem-solving activities, such as weather forecasting [35].

The proposed PSO-SVM model targets the expected exposure to periods of large temperature fluctuations. There are three units of the developed framework: the first of these is the so-called pre-processing unit, the second unit is a time forecast based on SVM, and the third unit is used to refine SVM parameters based on PSO. The conventional method's accurate prediction could not be given because the training set was not appropriate for prediction [36].

6.4.2 FFA-RF Model

It is possible to use an optimization algorithm to enhance the RF model. In order to fix optimization issues, numerous optimization algorithms have been implemented. Classical methods use a deterministic method to improve the optimum solution. The majority of evolutionary algorithms, under different complex conditions, are very nonlinear and multimodal. For optimization reasons, various AI optimization algorithms have been used, like swarm particle optimization (PSO), genetic algorithm (GA), artificial bee colony algorithm (ABC), and firefly algorithm (FFA). GA has a low rate of convergence, while there is a better convergence rate for FFA.

The FFA is a modern post-heuristic optimization widely used to overcome continuous multimodal optimization issues based on fireflies' behaviors. This study introduces a new hybrid FFA-RF algorithm that uses random forest approaches and firefly algorithms to predict solar radiation.

To build the suggested model, hourly weather information is used by finding the best number of trees and leaves per tree in the forest. Furthermore, the proposed methodology's effectiveness is compared to the traditional random forest model, the traditional ANN, and the FFA-optimized ANN model to demonstrate the superiority of the hybrid FFA-RF algorithm [37].

6.4.3   HHO-ANN

A new hybrid model focused on enhancing the efficiency of conventional artificial neural networks using the Harris Hawks Optimizer has been applied to predict the sun's functional productivity. This optimizer simulates Harris Hawk's activity to capture prey, and the optimal parameters for artificial neural networks are calculated by this method. Compared to two other models called SVM and Conventional ANN algorithms, the proposed methodology is called Harris Hawks Optimizer–Artificial Neural Network (HHO-ANN). The higher performance of HHO-ANN compared with ANN and SVM is demonstrated by four standard error parameters used to test data prediction performance: root mean square error (RMSE), determination coefficient (R^2), mean absolute percentage error (MAPE), and mean absolute error (MAE). Compared to the ANN algorithm, the proposed HHO-ANN provides better data prediction accuracy.

6.5   CONCLUSION

In this survey, different ML algorithms and novel hybrid models were analyzed for data prediction to improve wireless sensor networks' accuracy. We have discussed business and non-business issues in many aspects and the difference between WSN and machine learning strategies. Using this study, we can design an integrated framework that considers operational, non-operational, and specific applications to address the challenges of the WSN, i.e., a comprehensive framework for an energy-efficient WSN based on machine learning. This chapter has examined the machine learning techniques used in WSN in terms of data prediction. In application, ML algorithms are mainly used in information processing and data prediction, such as data processing and data transmission. This data prediction model is beneficial for wireless sensor network applications. We have to develop new novel optimization with a machine learning model from the literature survey to be developed for WSN data prediction.

ACKNOWLEDGMENTS

We would like to thank the Next Generation Internet of Things and Artificial Intelligence Laboratory at the Alagappa Chettiar Government College of Engineering and Technology, Karaikudi, for their support and advice in carrying out this research work.

REFERENCES

1. I. F. Akyildiz, W. Su, Y. Sankarasubramaniam, and E. Cayirci, "Wireless sensor networks: A survey." *Computer Network*, 38(4), 393–422, March 2002.
2. B. R. Stojkoska and K. Mahoski, *"Comparison of different data prediction methods for wireless sensor networks."* In *10th Conference for Informatics and Information Technology (CIIT 2013)*, Bitola, Macedonia, 2013.
3. U. Raza, A. Camerra, A. Murphy, T. Palpanas, and G. P. Picco, *"What does model-driven data acquisition really achieve in wireless sensor networks?"* In *Proceedings of the 10th IEEE International Conferences on Pervasive Computing and Communications*, 2012.
4. J. Cui and F. Valois, *"Data aggregation in wireless sensor networks: compressing or forecasting?"* In *Wireless Communication and Networking Conference (WCNC)*, IEEE 14.
5. R. Rajagopalan and P. K. Varshney, "Data-aggregation techniques in sensor networks: A survey." *IEEE Communications Surveys and Tutorials*, 4th quarter, 8(4), 48–63, 2006.
6. K. Akkaya, M. Demirbas, and R. S. Aygun, "The impact of data aggregation on the performance of wireless sensor networks." *Wireless Communications and Mobile Computing*, 8(2), 171–193, 2008.
7. Y. Liang and Y. Li, "An efficient and robust data compression algorithm in wireless sensor networks." *IEEE Communications Letters*, 18(3), 439–442, March 2014.
8. G. Krishnaa, S. K. Singhb, J. P. Singhb, and P. Kumar, *"Energy conservation through data prediction in wireless sensor networks"* in *3rd International Conference on Internet of Things and Connected Technologies (ICIoTCT)*, Jaipur, India, 2018.
9. B. Chander, B. Prem Kumar, and Kumaravelan, "A analysis of machine learning in wireless sensor network." *International Journal of Engineering & Technology*, 7(4.6), 185–192, 2018.
10. D. Barber, *Bayesian Reasoning and Machine Learning*, Cambridge University Press, New York, NY, 2012.
11. K. P. Murphy, *Machine Learning: A Probabilistic Perspective, MIT Press*, London, England, 2012.
12. A. Chlingaryan, S. Sukkarieh, and B. Whelan, "Machine learning approaches for crop yield prediction and nitrogen status estimation in precision agriculture: A review." *Computers and Electronics in Agriculture*, 151, 61, 2018.
13. M. A. Alsheikh, S. Lin, D. Niyato, and H. Tan, "Machine learning in wireless sensor networks: Algorithms, strategies, and applications." *IEEE Communications Surveys Tutorials*, 16(4), (1996–2018), Fourth quarter 2014.
14. T. Joachims, *"Text categorization with support vector machines: Learning with many relevant features"* in *European Conference on Machine Learning*, pp. 137–142. Springer, New York, 1998.
15. L. Breiman, "Random forests." *Machine Learning*, 45(1), 5, 2001.
16. C. Zhang and Y. Ma, *Ensemble Machine Learning: Methods and Applications*, Springer, New York, NY, 2012.
17. M. A. Rassam, A. Zainal, and M. A. Maarof," Principal component analysis-based data reduction model for wireless sensor networks." *International Journal of Ad Hoc and Ubiquitous Computing*, 18(1/2), 2015.
18. A. Rooshenas, H. R. Rabiee, A. Movaghar, and M. Yousof Naderi, *"Reducing the data transmission in wireless sensor networks using the principal component analysis"* in *Proceeding of the Sixth International Conference on Intelligent Sensors, Sensor Networks and Information Processing (ISSNIP '10) Brisbane*, QLD, Australia.

19. T. Yu, X. Wang, and A. Shami, "*A Novel R-PCA based multivariate fault-tolerant data aggregation algorithm in WSNs*," in *IEEE ICC 2016 Ad-hoc and Sensor Networking Symposium.*

20. J. Fattah, L. Ezzine, and Z. Aman, "Forecasting of demand using ARIMA model." *International Journal of Engineering Business Management*, 10, 1–9, 2018. doi:10.1177/1847979018808673.

21. R. Rebecca and S. Kalaivani, "Analysis of data prediction algorithms in wireless sensor networks." *International Journal for Research in Applied Science & Engineering Technology (IJRASET)*, 4, (V), ISSN: 2321–9653, May 2016.

22. C. Carvalho, D. G. Gomes, N. Agoulmine, and J. N. de Souza, "Improving prediction accuracy for WSN data reduction by applying multivariate spatio-temporal correlation." *Sensors*, 11, 10010–10037, 2011. www.mdpi.com/journal/sensors doi:10.3390/s111110010.

23. Y. Radhika and M. Shashi, "Atmospheric temperature prediction using support vector machines." *International Journal of Computer Theory and Engineering*, 1(1), ISSN: 1793–8201, April 2009.

24. Ch. Sai Sindhu, T. Hema Sai, Ch. Swathi, and S. Kishore Babu, "Predictive analytics using support vector machine." *International Journal for Modern Trends in Science and Technology*, 3 (Special Issue No. 2), March 2017, ISSN: 2455-3778.

25. M. S. Bennet Praba, A. J. Martin, S. Srivastava, and A. Rana, "Weather monitoring system and rainfall prediction using SVM algorithm." *International Journal of Research in Engineering, Science and Management*, 1(10), October 2018. www.ijresm.com. ISSN (Online): 2581-5792.

26. W. Zhang, H. Zhang, J. Liu, K. Li, D. Yang, and H. Tian, "Weather prediction with multiclass support vector machines in the fault detection of photovoltaic system." *IEEE/CAA Journal of Automatica Sinica*, 4(3), July 2017.

27. J. Lee, W. Wang, F. Harrou, and Y. Sun, "Reliable solar irradiance prediction using ensemble learning-based models: A comparative study." *Energy Conversion and Management*, 208, 112582, 2020.

28. S.-G. Kim, J.-Y. Jung, and M. K. Sim, "A two-step approach to solar power generation prediction based on weather data using machine learning." *Sustainability*, 11, 1501, 2019. doi:10.3390/su11051501.

29. W. Hameed, B. A. Sawadi, S. J. Al-Kamil, and M. S. Al-Radhi, "Prediction of solar irradiance based on artificial neural networks." *Inventions*, 4, 45, 2019. doi:10.3390/inventions4030045.

30. S. Samara and E. Natsheh," Intelligent real-time photovoltaic panel monitoring system using artificial neural networks." *IEEE Access*, 7(1), 50287–50299, 2019.

31. M. Chammas, A. Makhoul, and J. Demerjian, "An efficient data model for energy prediction using wireless sensors." *Computers and Electrical Engineering*, 76, 249–257, 2019.

32. X. Qing and Y. Niu, "Hourly day-ahead solar irradiance prediction using weather forecasts by LSTM." *Energy*, 148, 461e468, 2018.

33. H. Cheng, Z. Xie, L. Wu, Z. Yu, and R. Li, "Data prediction model in wireless sensor networks based on bidirectional LSTM." Cheng et al., *EURASIP Journal on Wireless Communications and Networking*, (2019), 203. doi:10.1186/s13638-019-1511-4.

34. J. Du, Y. Liu, Y. Yu, and W. Yan, "A prediction of precipitation data based on support vector machine and particle swarm optimization (PSO-SVM) algorithms." *Algorithms*, 10, 57, 2017. doi:10.3390/a10020057.

35. B. A. Selakov, D. Cvijetinović, L. Milović, S. Mellon, and D. Bekut, "Hybrid PSO–SVM method for short-term load forecasting during periods with significant temperature variations in city of Burbank." *Applied Soft Computing*, 16, 80–88, 2014.

36. I. A. Ibrahim and T. Khati, "A novel hybrid model for hourly global solar radiation prediction using random forests technique and firefly algorithm." *Energy Conversion and Management*, 138, 413–425, 2017.

37. F. A. Essa, M. Abd Elaziz, and A. H. Elsheikh, "An enhanced productivity prediction model of active solar still using artificial neural network and Harris Hawks optimizer." *Applied Thermal Engineering*, 2020. doi:10.1016/j.applthermaleng.2020.115020.

7 Strategic Sink Mobility Based on Particle Swarm Optimization in Wireless Sensor Network

Ramandeep Kaur and Roop Lal Sharma

St. Soldier Institute of Engineering and Technology, Jalandhar, Punjab, India

CONTENTS

7.1 INTRODUCTION

A wireless sensor network (WSN) is a network of numerous nodes (wireless in nature) that collect data from the network and forward the collected data to the data-collecting sink [1–3]. From the sink, it is sent to the user via the internet. There are a huge number of applications of WSN which deal with these two modes; static or mobile sink. Since the development of WSN, the data collection has been of major concern. Some of the researchers have used the static sink in the network, whereas some of them have exploited sink mobility in the network [4].

There are various advantages of sink mobility [5]. For applications where human reach is not possible, sink mobility plays an important role [6]. The primary concern of dealing with energy-efficient routing is to save the energy of the sensor nodes [7], because once the battery of the sensor nodes dies, we cannot replace it. Therefore, there is a high need to propose routing techniques which can reduce the energy consumption of sensor nodes so that these nodes can survive for a long period of time. The time unit in which the life of the sensor nodes is measured is termed as rounds. One round is said to be completed when the sink collects the data from all nodes in the network in one go. Therefore, the objective of the researchers working in the field of energy-efficient WSN is to increase the number of rounds in the WSN. Sink mobility is done with the help of a moving vehicle across the network or even inside the network. It is noted that the sink has no constraints of energy, computation, etc. Therefore, the concern is only about the nodes which are deployed in the network.

The question that arises is that of how to introduce sink mobility in the network. Various researchers have used different optimization methods for deciding about sink mobility. Recently, the particle swarm optimization (PSO) technique has been used to decide the sink mobility [8]. Kaur et al. in [9] proposed PSO-DSM (PSO-based dual sink mobility), in which the authors introduced the movement of two sinks in the network at the opposite corners, as shown in Figure 7.1. However, it is studied that the involvement of two sinks in the network increases the overall cost of the network, which sometimes becomes unaffordable for the user. The requirement of synchronization among the moving sinks is another concerning issue which needs careful consideration.

In other work, Sahoo et al. proposed the PSO-based energy-efficient clustering based on sink mobility (PSO-ECSM) protocol, in which the authors defined the sink mobility along with the cluster head selection [10]. In doing so the authors have increased the computational complexity of the network. Therefore, there is further scope for the improvement for the PSO-ECSM. Verma et al. presented GAOC that

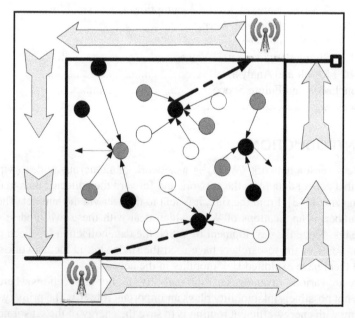

FIGURE 7.1   The scenario of PSO-DSM [9].

helped in the selection of CH by using the genetic algorithm (GA) [11]. The authors further introduced MS-GAOC (multiple-sink-based GAOC) by increasing the number of sinks in the network to four. However, the use of four sinks increased the cost of the network. Further, the sink is kept static in the network, which creates the complex scenario for the users. Verma et al. further proposed the MEEC protocol that used four gateways in the network to collect data from the network. With the use of four gateways, the need for synchronization is on the highly required side [12].

Therefore, there is a need to develop a PSO-based sink mobility protocol that considers the cost of the network as well the energy-efficient sink mobility.

7.1.1 CONTRIBUTIONS

Our major contributions in this work are stated as follows.

a. In this work, we have proposed strategic sink mobility in which the sink is made to move to the node which is having the least energy consumption rate, current energy level, and the distance of node. That particular node is identified through the optimization method using particle swarm optimization (PSO). The proposed scenario is shown in Figure 7.2.

b. The proposed technique is examined for its performance based on various performance metrics which are stability period, network lifetime, and many others.

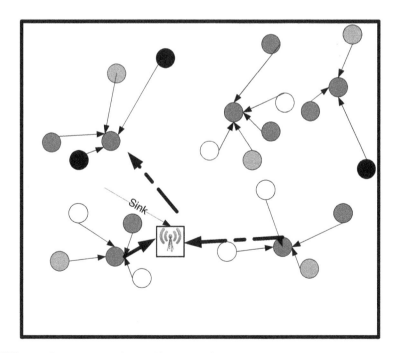

FIGURE 7.2 The proposed sink mobility scenario.

c. The comparison of the proposed technique is done with the recently developed methods namely, proposed work, PSO-DSM, and MEEC.

The rest of the chapter is structured as follows. In Section 7.2, we discuss related work in brief. In Section 7.3, the proposed framework is discussed. Further, we discuss simulation scenario in Section 7.4. The conclusion is given in Section 7.5.

7.2 RELATED WORK

WSN suffers from the energy limitation of sensor nodes. Therefore, to address this problem, various techniques have been proposed [13]. Ma et al. in [14] proposed an algorithm that considers the relay node fixation in the network for taking a control over the energy consumption. The local search approximation algorithm (LSAA) is used for handling this concern. Sensor nodes are distributed in the groups, and the LSAA algorithm helps in finding the time complexity along with the approximation ratio. The results show that proposed algorithm outperforms the RLSA algorithm in terms of saving the energy of the relay nodes. Saranya et al. in [15] proposed an algorithm, namely EECS, that decides about the network lifetime by the bits that are sent and the selection of CH. The CH is selected in a way that in the waiting time, the throughput can be maximized. Finite state machine (FSM) is idealized for a node having CH, CM, and different IDLE stages. The node which has the maximum residual energy, and is capable enough to send the greatest number of bits during the static interval of sink, is selected as CH. The Markov model is employed for the estimation of inter-state transitions. However, the algorithm has some limitations that cannot be afforded in the long-term survival of WSN. Only two parameters are used for CH selection, which is not sufficient for the optimal for any node.

Kumar and Kumar in [16] presented a LARCMS algorithm for routing that is location aware and that controls mobility of sinks. It is responsible for reducing the energy consumed by the nodes and lifetime improvement. Further, delay is minimized with the proposed strategy. Two mobile sinks are used in the network, and the routing technique performs better than the other algorithms. However, the cost of the network is enhanced and due to the involvement of two sinks, the number of overheads is enhanced dramatically.

Kaswan et al. in [17] proposed a design in which rendezvous points (RPs) decide for the path for efficient movement of the sink. For the other proposed methods, the delay bound factor is considered. To cover the whole operational area, the K-means clustering and other weight functions are considered. The data collection is handled with the scheduling method. The proposed algorithm is compared with many algorithms which are also working for a cluster-based mechanism, based on the simulation. The shortcoming is observed for the fact that a very small amount of time for data collection is given to MS and the operational nodes have a similar load for data generation.

Various other protocols have handled the energy consumption of the wireless sensor nodes pertaining to various applications [18–22]. However, it is observed that sink mobility has outperformed these methods to a great extent. The researchers' use of PSO-based techniques has proven to be significant in combating the energy consumption of the sensor nodes (Figure 7.3).

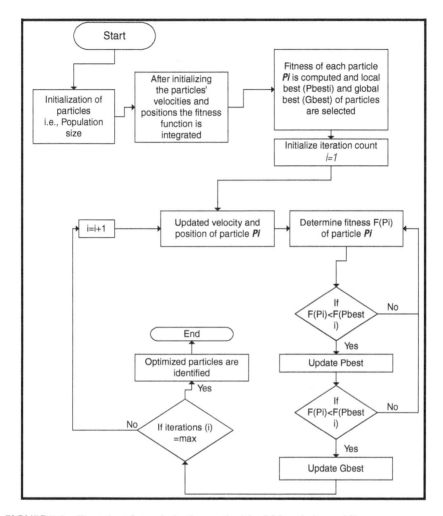

FIGURE 7.3 Flow chart for optimization method for PSO techniques. [4].

7.3 THE OPERATION OF PROPOSED WORK

In this section, we explain the operation of proposed work. The stepwise functioning of the proposed work is discussed as follows.

The algorithm-proposed work is triggered with a validation method in which the sensor nodes are given the identity of particles. Here are the following steps for the operation of the proposed work.

a. *Initialization*

Just after the validation process, the initialization is performed. The initialization of fitness function is similar to the initialization of parameters of the network.

b. *Design of the fitness function*

The computation of the fitness function is done with the combination of various parameters. These parameters are selected based on CH selection. The fitness factors that are used for designing this function are given as follows.

c. *The description of fitness parameters (FPs)*

The computation of FPs is done according to various factors. The energy optimization depends upon the extent of significance of these parameters. These parameters aim to abate the consumption of the energy and also are responsible for the elongation of the network lifetime. These parameters are considered for single objective function for the proposed work.

   i.   The node's residual energy

The residual energy of the node is the factor that defines the current value of energy [23]. It is important to know the status of the sensor node for a sink to move to any particular node. The sink is supposed to move to the node with the least energy first so that the data from that node can be collected.

FP_{1st} (First Fitness Parameter) is defined using Equation (7.1). As it could be seen, the energy factor is explored for this parameter.

$$F_{1st} = \sum_{i=1}^{N} E_{R(i)} \tag{7.1}$$

As given in Equation (7.1), the aggregated ratio of the ith node's residual energy given by $E_{R(i)}$.

   ii.   Distance factor considered for node sink

The second factor that is considered for optimization of sink movement is the measured distance. The sink will move to the sensor nodes nearest to it. While doing so, the sink can be made to move to the node at the earliest. The third fitness parameter (FP_{3rd}) meant for designing the fitness function for the sink movement is stated by Equation (7.2).

$$F_{2nd} = \sum_{i=1}^{N} D_{(N(i)-Sink)} \tag{7.2}$$

FP_{3rd} depicts the aggregated sum for the distance measured between the sensor node and data-collecting sink. Note that i varies from 1 to N (total nodes considered for simulation purpose).

   iii.   Energy Consumption Rate (ECR)

The last but most important factor that is considered while optimizing sink mobility, the ECR of every node is computed to check if the node is having optimum energy consumption rate. Only then is node considered for the next node for sink mobility. Equation (7.3) shows the third fitness parameter for ECR.

$$F_{3rd}\left(\text{ECR}\right) = \sum_{i=1}^{N}\left(E_{R(i)} - E_{AGG(i)} - E_{T(i)} - E_{Rx(i)}\right) \tag{7.3}$$

In Equation (7.3), the energy consumption rate is computed, $E_{AGG(i)}$ is the aggregation energy for the ith node

$E_{Rx(i)}$ is the energy consumed in the reception of data. $E_{T(i)}$ is the transmission energy of the ith node.

iv. Fitness function designed for the network

The fitness function of the network is the linear combination of three fitness parameters that are explained above as given in Equation (7.4).

$$F = \frac{1}{\varphi \times \text{FP}_{1st} + \delta \times \text{FP}_{2nd} + \gamma \times \text{FP}_{3rd}} \tag{7.4}$$

The fitness function denoted by F in Equation (7.4) needs to be reduced to acquire the optimized value of the performance of the network.

In Equation (7.5), φ, δ, and γ are the weight factors which are taken for representing each factor considered to design fitness function. For this work, the weight value to each factor is given through Equation (7.5).

$$\varphi + \delta + \gamma = 1 \tag{7.5}$$

7.3.1 SYSTEM CONSIDERATION OF PROPOSED WORK

The proposed network is subject to deployment of the heterogeneous nodes. These nodes are considered to give the scope for the CH selection of those nodes which could handle the energy consumption efficiently. There is a gradual decrease in the energy consumption of these nodes as the data transmission is commenced.

7.3.1.1 Network Model Assumptions Considered for Proposed Work

There are following network assumptions while designing a framework for the proposed work.

a. The network is considered to be stationary, and the sink is made to move in the network for collecting the data.
b. We consider nodes of various energy levels. These levels are normal, intermediate, and super nodes which are having different levels of energy.
c. There is no energy limitation on the sink.
d. The nodes are located at an unknown location.
e. The physical medium factors are not considered.
f. The shape of the network is considered to be square shape.
g. The distance computation among the nodes is done with the help of Euclidean distance which is further computed through Received Signal Strength Indicator (RSSI).

TABLE 7.1
Simulation Parameters for Proposed Work

Network Parameters	Values
Network dimensions	100×100 m²
Total number of nodes	100
Sink location	50, 50
Energy (in Joules) (E_o)	0.5
Level of heterogeneity	3 levels
Threshold distance (d_o)	87 m
Data aggregation	5 nJ/bit
Intermediate nodes energy fraction (β) and advanced nodes energy fraction (α)	$\beta = 1, \alpha = 2$
Intermediate nodes (m) and advanced nodes quantity fraction (m_o)	$m = 0.1, m_o = 0.2$
Simulation run	25
Total particles	30
Simulation run	20
Required energy for small distance $d \leq d_o$ (E_{efs})	10 pJ/bit/m²
Required energy for large distance $d > d_o$ (E_{mp})	0.0013 pJ/bit/m⁴
Packet size	2000 bits

7.4   SIMULATION SETTING SCENARIO

It is noted that the proposed work is based on simulation; hence, the MATLAB® software is given some network setting parameters. The MATLAB® application model 2016 is placed on a structure through a system with 2 GB RAM, 1 TB hard drive, Intel processor i3 by CPU consecutively on 3.07 GHz, in addition to Windows 7.

7.4.1   SIMULATION PARAMETERS VALUES

Table 7.1 mentions various values considered for the radio model and also the network dimensions. The deployment of the network is performed with 100 nodes that have different energy levels with different types of nodes.

7.4.2   RESULT AND ANALYSIS

We performed the simulation using the MATLAB® programming tool. Various performance metrics have been observed through the simulation outcomes of the proposed work. These metrics have been discussed individually in detail.

a. *Stability period*
   The stability period tells the rounds completed when the first node is dead. It is seen that proposed work has stability period equal to 8810 rounds, whereas the other protocols namely, PSO-DSM, MEEC, and GAOC, have less stability period, as shown in Figure 7.4. Therefore, the percentage improvement by the proposed work is 13% as compared to PSO-DSM, 12% and 13% as compared to MEEC, and 13% as compared to GAOC protocol, respectively.

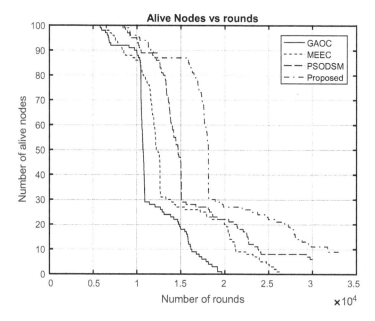

FIGURE 7.4   Alive nodes vs. rounds comparison of proposed work with other algorithms.

b. *Network lifetime*

When all nodes have no stock of energy left, it is defined as the network life-time of the sensor nodes. The proposed work is examined for this metric, as it is found the number of rounds covered are 33001 rounds, whereas the other protocols, PSO-DSM, MEEC, and GOAC, has network lifetimes equal to 30001, 26339, and 19649 rounds, respectively as shown in Figure 7.4. The percentage improvement is 10% as compared to the PSO-DSM protocol.

c. *Number of dead nodes against rounds*

This parameter gives the performance of the proposed work against the other protocols. In case of the proposed work, the first node dead and last node dead are found to be 8810 and 33000 rounds, whereas PSO-DSM has a smaller number of rounds for each case, as shown in Figure 7.5.

d. *Network's remaining energy*

This parameter is the metric which shows the pattern in which the energy is consumed by the sensor nodes. As the rounds are proceeded, the energy of the sensor nodes, which is 140 Joules, is consumed gradually. Figure 7.6 shows that the proposed work has fewer rounds for the particular value of rounds. The protocols PSO-DSM, MEEC, and GAOC cover fewer rounds as compared to the proposed work.

e. *Throughput/number of data packets sent to sink*

The number of data packets forwarded to the sink is shown in Figure 7.7. The proposed work shows the throughput with transmission of 1078336 data pack-ets, whereas PSO-DSM, MEEC and GAOC send 808932, 715605, and 698600 packets, respectively. The percentage improvement by the proposed work is

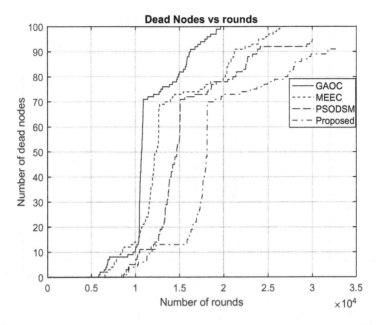

FIGURE 7.5 Comparison of dead nodes vs. rounds of proposed work with other algorithms.

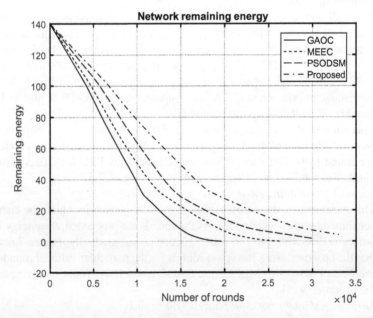

FIGURE 7.6 Network's remaining energy validation of proposed work with other algorithms.

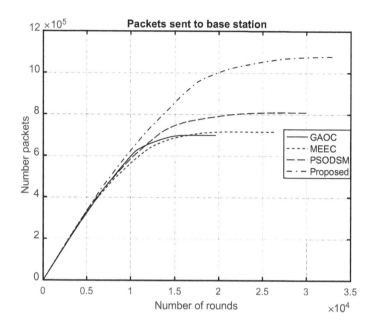

FIGURE 7.7   Comparison of throughput of proposed work with other algorithms.

33.3% as compared to the PSO-DSM protocol, 50.6% as compared to MEEC, and 54.3% as compared to GAOC. The improvement in the throughput is due to the optimized sink mobility in the WSN.

It is observed that in all of these metrics, the proposed work has outperformed the protocols GAOC, MEEC, and PSO-DSM. The reason behind such improvement is the consideration of the three crucial factors in the proposed work that uses the objective function of PSO. Whereas GAOC and MEEC consider static sink, the energy consumption for the proposed work is heavily reduced. Further, for PSO-DSM which uses dual sink mobility, the proposed work still works better than this due to the consideration of important parameters for CH selection. As a matter of fact, the proposed work is economical in the context of saving the financial cost for using one sink despite the use of two sinks in PSO-DSM.

7.5   CONCLUSION AND FUTURE SCOPE

The sensor nodes in the WSN are energy constrained. Therefore, there is a need for energy-saving techniques that can help in the network elongation. In this chapter, we use PSO-based sink mobility that helps in saving the energy of the nodes in the network. We have given the detailed description of algorithm and simulated over results using MATLAB® software. The sink is made to move strategically to an optimized node. The energy, distance, and energy consuming rate are considered for the sink movement. The simulation analysis shows the status for the performance of proposed technique over the various metrics namely, stability period, network lifetime, and throughput. We found that the proposed technique not only outperforms the existing methods but also gives a huge scope for the various applications where the network area is very large. In the future, we will focus on the energy harvesting of the sensor nodes so that the energy of the nodes can be preserved.

REFERENCES

1. H. Alemdar and C. Ersoy, "Wireless sensor networks for healthcare: A survey," *Computer Networks*, vol. 54, no. 15, pp. 2688–2710, 2010.
2. S. Verma, R. Mehta, D. Sharma, and K. Sharma, "Wireless sensor network and hierarchical routing protocols: A review," *International Journal of Computer Trends and Technology*, vol. 4, no. 8, pp. 2411–2416, 2013.
3. S. Verma and K. Sharma, "Zone divisional network with double cluster head for effective communication in WSN," *International Journal of Computer Trends and Technology*, vol. 4, no. 5, pp. 1020–1022, 2013.
4. B. M. Sahoo, H. M. Pandey, and T. Amgoth. "GAPSO-H: A hybrid approach towards optimizing the cluster based routing in wireless sensor network." *Swarm and Evolutionary Computation*, vol. 60, no. 2020, p. 100772.
5. B. M. Sahoo, A. D. Gupta, S. A. Yadav, and S. Gupta. "*ESRA: enhanced stable routing algorithm for heterogeneous wireless sensor networks.*" In *2019 International Conference on Automation, Computational and Technology Management (ICACTM)*, pp. 148–152. IEEE, 2019.
6. S. Verma, N. Sood, and A. K. Sharma, "Design of a novel routing architecture for harsh environment monitoring in heterogeneous WSN," *IET Wireless Sensor Systems*, vol. 8, no. 6, pp. 284–294, 2018.
7. C. Thomson, I. Wadhaj, Z. Tan, and A. Al-Dubai, "Mobility aware duty cycling algorithm (MADCAL) A dynamic communication threshold for mobile sink in wireless sensor network," *Sensors*, vol. 19, no. 22, p. 4930, 2019.
8. J. Kennedy et al., "Particle swarm optimization," *Encyclopedia of Machine Learning*, pp. 760–766, 2010.
9. S. Kaur and V. Grewal, "A novel approach for particle swarm optimization-based clustering with dual sink mobility in wireless sensor network," *International Journal of Communication Systems*, vol. 33, no. 16, p. e4553, 2020.
10. B. M. Sahoo, T. Amgoth, and H. M. Pandey, "Particle swarm optimization based energy efficient clustering and sink mobility in heterogeneous wireless sensor network," *Ad Hoc Networks*, p. 102237, 2020.
11. S. Verma, N. Sood, and A. K. Sharma, "Genetic algorithm-based optimized cluster head selection for single and multiple data sinks in heterogeneous wireless sensor network," *Applied Soft Computing*, vol. 85, p. 105788, 2019.
12. S. Verma, N. Sood, and A. K. Sharma, "A novelistic approach for energy efficient routing using single and multiple data sinks in heterogeneous wireless sensor network," *Peer-to-Peer Networking and Applications*, vol. 12, no. 5, pp. 1110–1136, 2019.
13. S. Verma and K. Sharma, "Energy efficient zone divided and energy balanced clustering routing protocol (EEZECR) in wireless sensor network," *Circuits and Systems: An International Journal (CSIJ)*, vol. 8, no. 6, pp. 284–294, 2014.
14. C. Ma, W. Liang, M. Zheng, and H. Sharif, "A connectivity-aware approximation algorithm for relay node placement in wireless sensor networks," *IEEE Sensors Journal*, vol. 16, no. 2, pp. 515–528, 2015.

15. V. Saranya, S. Shankar, and G. R. Kanagachidambaresan, "Energy efficient clustering scheme (EECS) for wireless sensor network with mobile sink," *Wireless Personal Communications*, vol. 100, no. 4, pp. 1553–1567, 2018.

16. V. Kumar and A. Kumar, "Improving reporting delay and lifetime of a WSN using controlled mobile sinks," *Journal of Ambient Intelligence and Humanized Computing*, vol. 10, pp. 1–9, 2018.

17. A. Kaswan, K. Nitesh, and P. K. Jana, "Energy efficient path selection for mobile sink and data gathering in wireless sensor networks," *AEU – International Journal of Electronics and Communications*, vol. 73, pp. 110–118, 2017.

18. S. Verma, N. Sood, and A. K. Sharma, "QoS provisioning-based routing protocols using multiple data sink in IoT-based WSN," *Modern Physics Letters A*, vol. 34, no. 29, p. 1950235, 2019.

19. S. Verma, S. Kaur, A. K. Sharma, A. Kathuria, and M. J. Piran, "Dual sink-based optimized sensing for intelligent transportation systems," *IEEE Sensors Journal*, 2020. doi:10.1109/JSEN.2020.3012478.

20. Sahoo, Biswa Mohan, Ranjeet Kumar Rout, Saiyed Umer, and Hari Mohan Pandey. "*ANT colony optimization based optimal path selection and data gathering in WSN.*" In *2020 International Conference on Computation, Automation and Knowledge aq Management (ICCAKM)*, pp. 113–119. IEEE, 2020.

21. S. Verma, N. Sood, and A. K. Sharma, "Cost-effective cluster-based energy efficient routing for green wireless sensor network," *Recent Advances in Computer Science and Communications*, vol. 12, pp. 1040–1050, 2020.

22. S. Verma, S. Kaur, R. Manchanda, and D. Pant, "*Essence of blockchain technology in wireless sensor network: a brief study.*" In *2020 International Conference on Advances in Computing, Communication & Materials (ICACCM)*, pp. 394–398, 2020.

23. S. Verma, S. Kaur, M. A. Khan and P. S. Sehdev, "Towards green communication in 6g-enabled massive internet of things." In *IEEE Internet of Things Journal*. doi:10.1109/ JIOT.2020.3038804.

8 A Study on Outlier Detection Techniques for Wireless Sensor Network with CNN Approach

Biswaranjan Sarangi

Biju Patnaik University of Technology, Odisha, India

Biswajit Tripathy

GITA, Bhubaneswar, Odisha, India

CONTENTS

DOI: 10.1201/9781003145028-8

8.1 INTRODUCTION: WIRELESS SENSOR NETWORKS (WSN)

A wireless sensor network is a bridge between the virtual world of information technology and the real physical world. A sensor node is essentially the latest trend of Moore's Law toward the miniaturization and ubiquity of computing devices. Typically, a wireless sensor node (or simply sensor node) consists of sensing, computing, communication, actuation, and power components. These components are integrated on a single board or on multiple boards, and packaged in a few cubic inches. A Berkeley Moto, perhaps the first sensor device developed under LWIM (low-power wireless integrated microsensors) project at University of California, Los Angeles (UCLA), was funded by DARPA. The large number of these low-power, inexpensive sensor devices are densely embedded in the physical environment, operating together in a wireless network and referred to as a wireless sensor network (WSN). Figure 8.1 shows the key components of a WSN device.

A 8 bit 16 MHZ low power embedded processor performs both the computational task of locally sensed information and information communicated by other sensors. The memory or storage includes both program memory and data memory in the form of random access. The radio transceiver having a low-rate (10–100 Kbs) and short-range (< 100 m) wireless radio having energy-efficient sleep and wake-up modes. Low-data-rate sensors on board, depending on application, include temperature sensor, pressure sensor, heat sensor, humidity sensors, etc. Pre-configured sensor locations at deployment can be obtained by satellite-based GPS for positioning. The power source by LiMH AA batteries of WSN devices for finite energy is a measure challenge in most WSN applications.

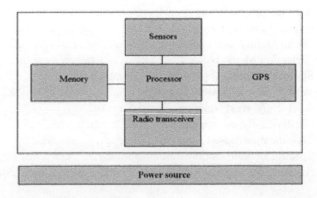

FIGURE 8.1   Block diagram of WSN.

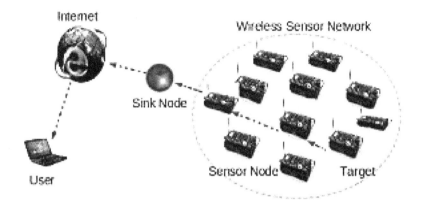

FIGURE 8.2   Architecture of wireless sensor network.

The two objectives of network deployment of these WSN devices are coverage (application-specific quality of information) and connectivity (which pertains to the network topology over which information routing can take place). There are different types of deployment strategies as structured versus randomized deployment, over-deployment versus incremental deployment, network topology like star, grid, hierarchical, etc., homogeneous versus heterogeneous, and coverage metrics.

In the context of network topology, every node communicates its measurement directly to the gateway in a single-hop star topology. Multi-hop mesh and grid depends on how they are placed using arbitrary mesh graph or 2D grid. In a two-tier hierarchical cluster, it naturally decomposes a large network into separate zones within which data processing and aggregation can be performed locally. Figure 8.2 shows the architecture of the WSN.

8.1.1 Application of WSN

WSNs are used in essential applications such as remote patient health monitoring, environmental monitoring, engineering structure monitoring, industrial and commercial networked sensing, military detection and goal tracking, and so on. Scientific studies of animals, plants, and micro-organisms, referred to as observer effect, are conducted by deployment of wireless sensor networks. The sensor network transmits the data over the web via a satellite communication link. In military surveillance and target tracking applications, the sensor nodes are deployed rapidly to get the information related to location, numbers, movements, and identity of troops and vehicles and also for detection of chemical, biological, and nuclear weapons. In case of structural and seismic monitoring applications, sensor networks are used for monitoring the long-term wear of structures and their conditions after earthquake or explosion types of events. In industrial and commercial networked sensing, sensors and actuators are used for process monitoring and control.

8.1.2 WSN Design Challenges

Sensor nodes have the energy constraint because of low battery power, which works less than a month if it is operated continuously in full active mode. In some unattended cases, we need the nodes to be operational for years without changing the battery. Energy harvesting technique and hardware improvement in battery design provide a partial solution. Responsiveness between sleep and wakeup modes of nodes by synchronization is a challenging task. To provide robustness, the protocol design must have a built-in mechanism when the nodes are deployed in harsh and hostile environment prone to failure. Designing a synergistic protocol which can provide an efficient collaborative use of storage, computation, and communication resources is a measure challenge. In some applications, the nodes are inherently unattended, and distributed and autonomous operation of the network like self-configuration is a major design challenge.

8.2 OUTLIERS

Recently, outlier detection is of higher interest to researchers and has attracted much attention for several real-life applications. The information must be accurate and complete, and this is generated by the nodes. The data collected from these sensor nodes is analyzed in a timely manner. Sometimes the raw data generated by the nodes are inaccurate and incomplete when battery power is low and due to suspicious environmental effects. So it is necessary to identify these erroneous data occurring in the network [1]. Figure 8.3 shows outliers in a two-dimensional data set.

8.2.1 Definitions of Outliers

An outlier is an observation which is suspected of being partially or wholly irrelevant because it is not generated by the stochastic model.

[2–4]

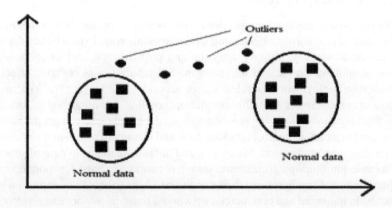

FIGURE 8.3 Outliers in a two-dimensional data set.

An outlying observation, or outlier, is one that appears to deviate markedly from other members of the sample in which it occurs.

[2]

An outlier is an observation that deviates so much from other observations as to arouse suspicion that it was generated by a different mechanism.

[2]

An outlier is an observation or subset of observations that appears to be inconsistent with the rest of the set of data.

[3]

Outliers are points that lie in the lower local density with respect to the density of its local neighborhood.

[4]

Outliers are points that do not belong to clusters of a data set or as clusters that are significantly smaller than other clusters.

Points that are not reproduced well at the output layer with high reconstruction error are considered as outlier.

A point can be considered as an outlier if its own density is relatively lower than its nearby high density pattern cluster, or its own density is relatively higher than its nearby low density pattern regularly.

If the removal of a point from the time sequence results in a sequence that can be represented more briefly than the original one, then the point is an outlier.

A point is considered to be an outlier if in some lower-dimensional projection it is present in a local region of abnormal low density.

A spatial-temporal point whose non-spatial attribute values are significantly different from those of other spatially and temporally referenced points in its spatial or/and temporal neighborhoods is considered as a spatial-temporal outlier.

An outlier is a data point which is significantly different from other data points, or does not confirm to the expected normal behavior, or conforms well to a different abnormal behavior.

Those measurements that significantly deviate from the normal pattern of sensed data.

[4]

8.2.2 Types of Outliers

There are two types of outliers such as local and global outliers depending on the scope of data. Local outliers, referred to as first order outliers, are those where the density of an object in a data set deviates from the local area, leading to inconsistent observation in comparison to the rest of the data.

Higher order outliers, also known as global outliers, are determined by the network's design, in which the whole data set is employed as background detection and detected at various levels [5]. In a centralized architecture, all data are transmitted to the sink node. For outlier detection, the data are collected within its controlling range through an aggregator in an aggregated architecture.

8.2.3 Sources of Outliers

The main sources of outliers are events, malicious attack, noise, and error. From a faulty sensor, a noise-related measurement is considered as an outlier, where outliers by error occur recurrently. Outliers by events, like forest fire, air pollution, etc., have a lower probability of incident which lasts for a long time and changes the historical pattern of data [6]. Figure 8.4 shows outlier sources in WSNs.

8.2.4 Degree of Being an Outlier

Scalar and outlier score are two measurements of outlier [7]. In scalar scale, each measure of data is classified into normal or outlier class known as binary classification measure. In outlier score measurement, the outlier score is assigned to each data measurement depending on the degree and provides a ranked list of outlier. We can choose the top n outliers from the ranked list to analyze, or we can choose a cutoff threshold for outlier measurement.

8.2.5 Dimension of Outliers

Sensor data is viewed as a stream of data which can be univariate or multivariate based on the set of values coming from a particular sensor node or values from different sensors of the same sensor node.

8.2.6 Data Correlation

In both time and space, sensor data appears to be associated [8]. Changes in data values over time cause temporal similarity at a single node position. It denotes the association between the data at timestamp t and the data at its background node at previous time instants. Due to comparisons with adjacent nodes, spatial similarity exists at a single node position. It means that the data of sensor nodes within a certain physical spatial context, such as neighborhood and cluster, have a geographical connection. Due to shifts in data significance over time and space, spatiotemporal similarity exists across a variety of node positions.

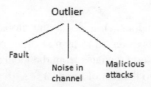

FIGURE 8.4   Outlier sources in WSNs.

8.2.7 ARCHITECTURAL STRUCTURE

The data collected through sensor nodes are sent to base station to be processed and analyzed in centralized architectural approach. The outlier detection is performed in the sink node having more resources and large storage capacity. A clustering technique is used for higher detection accuracy. However, outlier detection is performed at each sensor node in distributed architectural approach through real-time detection methods.

8.2.8 ISSUES OF OUTLIER DETECTION

The main challenge of outlier detection is to achieve high detection accuracy with minimum energy cost and maximum effectiveness. To design a suitable solution, the major constraints need to be analyzed by the following characteristics.

Source restraints. The computational capacity, power, communication bandwidth, and storage capacity are major resources for process of outlier detection in a WSN. But the uses of cheap and low-quality sensors are major constraints in terms of resources. Extreme route failure, signal absorption, scattering, quickly shifting times-varying channels, wide propagation latency, noise, and fading characteristics are all significant contact restrictions in WSN. Data transmission costs are greater than data production costs. The amount of connectivity overhead gained by the delivery method has an effect on the energy usage of distributed online outlier detection models.

Distributed streaming data. The transient existence of sensor data poses a significant challenge since its distribution may alter over time, and there is no prior information to build a normal reference model distribution of sensor data.

Dynamic network topology. In a harsh and unattended environment, when the sensor nodes are deployed over a long period of time, the network topology is susceptible for frequent communication failure. Some new nodes are to be added, and some defective nodes are to be deleted to the network; as a result, the old normal reference model needs updating.

Network heterogeneity. Sensor nodes having different type of sensors for measuring different environmental phenomena at the same time refers to network heterogeneity. The data collected by different sensors follow different data distribution techniques, making the outlier detection model incapable.

Identifying outlier sources. Due to the complex existence of WSNs and resource limits, determining the origin of a network outlier is challenging. A fundamental difficulty is collecting raw data from sensors to be processed in a decentralized and online way while keeping memory, connection overhead, and computing cost low.

High-dimension data. The dimensionality of collected data increases as the network size increases, which incurs high computational cost and reduces the power source of the sensor, as well as drains the memory. Also, the high-dimension data becomes a major problem for efficient outlier detection.

8.2.9　Use of Outlier Detection in WSN

Dealing with outliers depends upon the application domain. The application areas are so diverse that it is not possible to cover all, and a few application areas of interest are considered.

Habitat monitoring. The endangered species are equipped with small nonintrusive sensors to monitor their behavior.

Environmental monitoring. Different types of sensors are deployed in harsh and hostile environment to measure humidity, temperature, etc., to monitor the nature of environment.

Healthcare and medical monitoring. Small sensors are connected to the body of the patient to monitor vital status and patient metrics such as heart rate, pulse rate, and blood pressure.

Target tracking. Sensors are embedded in moving objects to track them at a particular time.

Industrial monitoring. Devices are equipped with pressure, temperature, or vibration sensors to monitor their states.

Internet of Things. IoT devices are made of a lot of sensors that sense environmental parameters depending on the desired task. It is essential to check the quality of data before carrying out of the task, as it may be corrupted with outliers. Detection of these outlier data leads the overall efficiency.

Time-series monitoring and data streams. Detecting the outlier in time-series data generated by sensors at different time spans and detecting the abnormal pattern in data streaming is of high importance, as it may influence the fast computation and estimation of correct output.

Data quality and data cleaning. Data generated from defective sensors sometimes may contain faulty data that need to be cleaned, as having the correct data is essential for training high-quality model for prediction of accurate results.

8.3　OUTLIER DETECTION METHODS

In this section we describe different outlier detection techniques used for WSN based on different approaches, as shown in Figure 8.5.

8.3.1　Statistical-Based Approach

The statistical approach is based on a probability distribution model for a given data set where the model evaluate the data instances to fit with and a lower probability of data instances leads to an outlier. There are two techniques, parametric and non-parametric.

a. The parametric technique takes into account the primary data distribution and estimates the distribution parameters from the available data. Based on the form of distribution, these methods are further classified into Gaussian-based and non-Gaussian-based models.

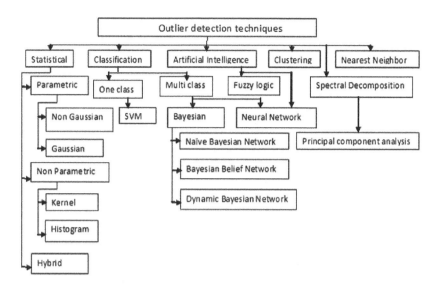

FIGURE 8.5 Outlier detection techniques.

- In a Gaussian-based model, normal distributions of data are presumed, and two statistical tests are used to find outliers locally [9–11], which deal with one-dimensional outlier data and take so much memory for a node to hold old values.

- The data in a non-Gaussian-based model are not usually distributed [12]. It finds outliers in the form of impulsive noise using a symmetric-symmetric table (s × s) distribution that is not optimal for actual sensor data. The sensor data spatial-temporal correlation is used to detect outliers locally. It decreases connectivity costs due to local propagation, as well as computing costs since the cluster-heads do the bulk of the computation. Hybrid outlier detection technique is a semi-supervised, local outlier detection method. It identifies errors and detects events in ecological applications of wireless sensor networks.

b. In non-parametric methodology, the distance measure notions between a new test instance and the statistical model are established. This procedure does not depend on the availability of data delivery and therefore uses a threshold to define an outlier using two techniques. The first method is focused on calculating density first and then classifying it, while the second method is based on selecting a group explicitly. To extract data distribution from the network and filter out the non-outliers, the sink node uses histogram information for centralized processing [13].

8.3.1.1   Kernel-Based Approach

In this approach, a kernel density estimator is used for outlier identification online [14,15]. If the number of values in its neighborhood is less than a threshold value which is user specified, then this value is treated as an outlier. This approach does not detect spatial outliers, and the high reliance on the defined threshold is the main drawback.

8.3.1.2 Nearest Neighbor-Based Approach

Several distance notions like Euclidean distance are used to analyze the data, and if the data instance is to be found far from its neighbor, then the data is treated as outlier data. This approach is generally used in data mining and machine learning. Use of suitable input parameter is the main constraint.

8.3.2 CLUSTERING-BASED APPROACH

Grouping the data into classes is known as clustering, where cluster objects have similarity between them and dissimilarity to objects within other clusters. Here, the set is partitioned into groups based on similar data and labels are assigned to small number of groups. This approach is generally used in unsupervised learning and suitable for outlier detection of temporal data [16]. A suitable cluster width parameter is highly required in this approach.

8.3.3 CLASSIFICATION-BASED APPROACH

First, we have to learn a classification model using data set instances. Then this is classified as training class, where prior knowledge of data set is not essential.

The support vector machine-based classification method is generally used for both linear and non-linear data. The original training data is transformed to a higher dimension data by a non-linear mapping, and in this high-dimension data, a linear hyper plane is searched so that it can separate the tuples from one class to another. The support vector machine finds this hyper plane using margins and support vector. During the training phase it learns a classification model, and this model is later used to classify any new arrived or unseen or unobserved data. Euclidian distance metrics are used in hyper plane-, hyper sphere-, and quarter sphere-based SVM where as Mahalanobis distance metrics are used in hyper ellipsoidal, centered ellipsoidal SVM to detect outlier. Hyper plane SVM is not ideal for power-constrained WSNs in rugged environments due to its low classification and generalization abilities. Although the hyper sphere SVM has strong generalization capability, it is not feasible to apply on energy-constrained WSNs due to a quadratic optimization challenge. The quarter sphere SVM methodology takes into consideration the spatial temporal association of sensor nodes, rendering it capable of managing both local and global outliers [17]. In comparison to the hyper sphere SVM, the hyper ellipsoid SVM has a quadratic optimization problem and needs more computing and memory resources. The centered ellipsoid SVM method takes into account multivariate and streaming results, as well as spatial and temporal correlations. The support vector data description, based on the spatiotemporal and attribute correlations (STASVDD)-based approach, is suggested in [18] to detect outliers; this considers that outliers can independently occur in each attribute when the collected data vectors are independent and identically distributed in WSNs.

The Bayesian method is focused on three distinct forms of probabilistic regression:

- *Naive-Bayesian network model.* The meaning of an attribute on a specified class is assumed to be independent of the values of other attributes by Bayesian classifiers. This statement is regarded as class conditional independence, and it is called naive since it is made to simplify the calculation. It is called an outlier if the likelihood of a sensed reading in its class is lower than the probability of being in other groups [19].

• *Bayesian Belief Network.* This network is a graphical representation of a probabilistic dependence model made up of interconnected nodes. In the dependency model, each node represents a vector. The linking arcs reflect the causal interactions between these variables. Two elements make up a conviction network. A guided acyclic graph is one in which each node represents a random variable and each arc represents a probabilistic dependency. For each vector, the second part of a belief network is a conditional probability panel. Learning the network is simple if the network function is understood and the variables are measurable [8].

• *Dynamic Bayesian Network.* The Bayesian networks reflect the vector sequences in this case. Influence diagrams are generalizations of Bayesian networks that describe and solve decision problems under uncertainty. DBNs are used in this methodology to detect changes in sensor network topology quickly. Outliers are identified using this method, which calculates the posterior likelihood of the most recent data values in a sliding window. Outliers are data measures that fall outside of the predicted value interval [20]. This method will immediately manage several data sources.

8.3.4 SPECTRAL DECOMPOSITION-BASED APPROACH

The aim of this method is to use principal components analysis to find typical modes of action in data by restricting multidimensional data sets to lower-dimension data sets. This data analysis is generally used for making predictive models by calculating eigenvalue decomposition or singular value decomposition in a data set [21].

8.3.5 ARTIFICIAL INTELLIGENCE-BASED APPROACH

Recently, more attention has been given in learning-based methods such as active learning and deep learning for outlier detection problems in WSNs [22]. Convolutional neural network (CNN), stacked autoencoders (SAE), deep belief networks (DBN), long short-term memory recurrent networks (LSTM), etc., are several deep learning neural networks which are well suited for dealing with different classification problems. These neural networks deal with nonlinear large-scale data with skewed properties [23]. The deep learning method is a good choice in the field of outlier detection because of the advantages of learning the features directly from the original data automatically.

In [24], the autoencoder neural networks is used to solve the anomaly detection problem in WSN. The authors have developed a two-part algorithm, which resides respectively on sensors and the IoT cloud. The anomalies are detected in a distributed manner at sensor nodes without having to communicate with any other sensor nodes or the cloud. The authors of [25] have proposed two solutions to outlier detection in time series based on recurrent autoencoder ensembles. The solutions exploit autoencoders built using sparsely connected recurrent neural networks (S-RNNs), which make it possible to generate multiple autoencoders with different neural network connection structures. Both solutions, in particular an independent framework and a shared framework, combine multiple S-RNN-based autoencoders to allow outlier detection.

8.4 OUTLIER DETECTION USING CNN

We attempted to convert the sensor data streams generated by sensor nodes which are densely deployed over a large geographical area into images and used a CNN model for learning and testing. A CNN model recognizes complex function approximation by learning the deep nonlinear network structure. It represents the input-output mapping relationship and simultaneously learns the basic characteristic of a data set from a small sample set.

The convolutional neural network (CNN) is a multilayer neural network named from a mathematical linear operation between matrixes, known as convolution. It has four main layers: convolutional layer, *ReLu* layer (nonlinearity), pooling layer, and fully connected layer. CNN architecture is formed by a stack of these layers. The convolutional and fully connected layers have parameters, but the pooling and nonlinearity layer do not [26]. Figure 8.5 shows the outlier detection process using CNN.

1. *Convolutional layer.* The convolutional layer plays a vital role in CNN operation. The parameters of this layer emphasize the use of learnable kernels which consist of several feature maps. It determines the output of neurons which are connected to local regions of the input through the scalar product calculation between their weights for each value in that kernel. Every kernel having an activation map stacked along the dimensional depth forms the output. When the data strike the convolutional layer, the layer convolves each filter of the input produces a 2D activation map.

2. *ReLu layer.* The rectified linear unit, commonly referred as ReLu, is a nonlinear activation function, also referred to as piecewise linear function. It is chosen for calculating the feature map generated by the filters. The feature map is calculated by the ReLu activation as

$$h_i^k = \max\left(W^k x_i, 0\right) \tag{8.1}$$

where
  h^k is the kth feature map at a given layer
  i is the feature map index
  x_i is the input
  w^k denotes weights

Because of the computational simplicity, representational sparsity, and linear behavior, the ReLu activation function has become the default activation function used in almost all convolutional networks.

3. *Pooling layer.* The main concept of this layer is to reduce the dimension of the feature maps which reduce the complexity for further layers and increases the

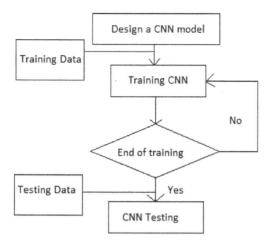

FIGURE 8.6   Outlier detection process using CNN.

robustness of feature extraction. The pooling layer does not affect the number of filters and performs down sampling along the spatial dimensionality of the given input. Maxpooling is considered as the default pooling layer which reduces the number of parameters within that activation.

$$f_{max}(x) = max(x_i) \qquad (8.2)$$

where x is input data vector with activation values.

4. *Fully connected layer.* After all of the features are generated, they are passed to the softmax fully connected layer. Each node in this layer is directly connected to every node in both the previous and the next layer by taking all the neurons from the previous layer and combining them into one layer. The fully connected layer contains neurons which are directly connected to the neurons in two adjacent layers, without being connected to any layers within them. The output of the fully connected layer is the probability distribution of all classes which is the final result of classification. Figure 8.6 shows the outlier detection process using CNN.

8.4.1   PROPOSED APPROACH

The data generated from different sensor nodes deployed at different locations are collected with a specific epoch duration and stored in a data set. The data preprocessing is to convert the data in the data set to time series data and then outlier detection of this data by using the CNN model with EEG classification. We proposed convolutional neural network which can deal with the structure of EEG data, as it is able to learn two-dimensional representation of data. The proposed EEG classification architecture is shown in Figure 8.7, where sensor data is converted to time series and spectral power within theta, alpha, and beta frequency is extracted for each time slice

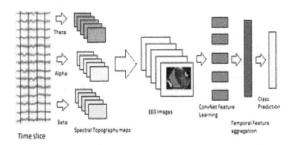

FIGURE 8.7   EEG classification architecture.

and used to form sequential topographical maps for each time frame and combined to form a three-channel image used in CNN for classification and learning.

In this chapter we propose the methodology consisting of the following five steps.

Step 1. In data processing, the data were separated into overlapping one-second frames. Windowing is the process of taking a small subset of a larger data set, for processing and analysis which alter the spectral properties of that data set. Here a Hann window is applied over the frames. The Hann function, named after Julius Von Hann, is used to create a "window" for Fourier Transform filtering. The Hann window is defined as

$$w(n) = \alpha - \alpha * \cos\left(\frac{2\pi n}{M-1}\right), 0 \le n \le M-1. \text{Where } \alpha = 0.5. \qquad (8.3)$$

The factor 1/M is chosen rather than 1/M − 1 to give the best behavior for spectral estimation of discrete data as the end points of Hann value just touch 0. It is also known as raised cosine, as it is a member of cosine-sum and power-of-sine families.

Step 2. To transform the data of each frame from time domain to frequency domain, First Fourier transform (FFT) is then applied. Fourier transform is a function which transforms a time domain signal into the frequency domain by accepting a time signal as input and produces the frequency representation as an output [27]. Looking at signals in the frequency domain can help for validating and troubleshooting the signals. The frequency domain is great at showing if a clean signal in the time domain actually contains noise or jitter.

Step 3. FFT amplitudes are classified into theta (4–8 Hz), alpha (8–12 Hz), and beta (12–40 Hz) ranges giving three scalar values for each probe per frame. Real-world data tend to be noisy, so data cleaning routines attempt to smooth out noise while identifying outliers in the data. Here we are concerned with frequency binning for data smoothing. In this method the data is first sorted and then distributed into a number of buckets or bins. The binning approach smooths sorted data values by conferring with neighbor data, that is, the value surrounding it. In equal depth or frequency binning, we divide the range of the variable into intervals that contain approximately an equal number of points, and equal frequency may not possible due to repeated values. Binning is one of a number of different data smoothing techniques. Smoothing by "bin means" is where each value in a bin is replaced by the mean value of the bin. Smoothing by "bin median" is where each value in a bin is replaced by the median value of the bin. Smoothing by "bin boundary" is where the minimum and maximum values in a given bin are identified as the bin boundaries and then each bin value is replaced by the closest boundary value.

Step 4. The three scalar values generated from Step 3 are converted as RGB color channels onto a 2D map by 2D Azimuthal projection. Preservation of the directions from a central point is the property of Azimuthal projection, where the straight lines on the map represent great circles through the central point. These projections have radial symmetry in the scales and also in distortions. A plane tangent to the map among the central point as a tangent point can be imagined and visualized as the mapping of radial lines.

Whether the plane is tangent (the plane having one point of contact) or secant (the plane having an entire line of intersection), we can minimize the level of choosing standard lines. The three values of theta, alpha, and beta were converted as RGB color channels onto a 2D map projection. The function computes the Azimuthal equidistant projection of input point in 3D Cartesian coordinates, and the result returns projected coordinates using Azimuthal equidistant projection.

Step 5. These 2D map projections of theta, alpha, and beta ranges are trained in CNN to measure the validation accuracy and detection accuracy.

In this study, we analyzed some data samples extracted from the data set [28] of the sensor-scope-Lausanne Urban Canopy Experiment (LUCE), which was collected between July 2006 and May 2007 through a sensor scope project on the campus of the Ecole Polytechnique Fédérale de Lausanne (EFPL). In the central part of the campus, a network of 92 wireless weather sensor nodes covering an area of 300 × 400 m was deployed. The sensor nodes were deployed to measure ambient temperature, surface temperature, humidity, wind speed, etc. From the sensor, nodes with ID 10 are included in this work by taking data based on time. Synthetic outliers are generated by using fault models suggested in [29] and inserted into the data set.

The data in the dataset converted to waveform data and the waveform signals from 17 days with time slices are taken with a sampling rate of 128 Hz.

8.4.2 EXPERIMENTAL SETUP

A high-level python library, Pymote 2.0 [30], is used for simulation and Tensorflow to implement our CNN model. Seventy percent of the data set was used for training, and 30% was used for testing. A fully connected layer and a softmax layer preceded by a VGG-style CNN network is used. A maxpool layer separates the stack, and the kernels number in each layer becomes double of previous stack. The complete network with maxpooling layer in time was built by inputting the EEG images (one image per time window), number of classes, size of the input images (a square input), number of color channels (three—RGB), and number of time window in the snippet returns a pointer to the output of the last layer. The conv2D layer builds the complete network to integrate time from sequence of EEG images.

A sample training function was built which loops over the training set and evaluates the network on the validation set after each epoch. It evaluates the network on the training set by inputting the images, target labels, tuple of (train, test) index numbers, model type, batch size for training, and number of epochs of data set to go over for training.

8.4.3 EVALUATION METRIC

The following metrics are used to evaluate the performance of the proposed method.

$$\text{Accuracy rate} = \frac{TP + TN}{TP + TN + FP + FN} \tag{8.4}$$

$$\text{Precision}(P) = \frac{TP}{TP + FP} \tag{8.5}$$

$$\text{True Positive Rate}(\text{Recall})\,TPR = \frac{TP}{TP + FN} \tag{8.6}$$

$$\text{False Positive Rate}\,FPR = \frac{FP}{FP + TN} \tag{8.7}$$

$$F1 = \frac{2(Precision \times Recall)}{Precision + Recall} \tag{8.8}$$

In Equation 8.4, TP denotes the number of true-positive results, TN represents the number of true-negative results, FP is the false-positive results, and FN represents the number of false-negative results. In Equation 8.8, F1 is the harmonic mean that measures the quality of classifications.

8.5 CONCLUSION

The main objective of outlier detection is to identify the unruly nodes and to restrict the data reported by those nodes to enter into the network. In this chapter we have presented a CNN-based online outlier detection method that is integrated with EEG classification. This approach transforms the sensor data into sequence of EEG images in the preprocessing step. The performance according to accuracy, TPR, FPR, precision, and F1 is compared with the state-of-art techniques which show significant improvements in detection accuracy as shown in Table 8.1.

TABLE 8.1
Comparison with Experimental Results

Method	AR	TPR	FPR	P	F1
SVM [17]	82.29	84.86	42.11	95.05	89.66
DNDO [31]	84.51	87.04	39.83	95.44	91.05
Proposed	85.21	89.45	27.10	95.62	92.54

REFERENCES

1. A. Mahapatro and P. Khilar, Fault diagnosis in wireless sensor networks: A survey, *Communications Surveys Tutorials IEEE*, vol. 15, no. 4, pp. 2000–2026, 2013.
2. D. M. Hawkins, *Identification of Outliers*, London: Chapman and Hall, 1980.
3. V. Barnett and T. Lewis, *Outliers in Statistical Data*, New York: John Wiley & Sons, 1994.
4. S. Sadik and L. Gruenwald, *Online outlier detection for data streams*, in *Proceedings of the 15th Symposium on International Database Engineering and Application*, ACM, Lisbon, Portugal, pp. 88–96, 2011.
5. V. Chatzigiannakis, S. Papavassiliou, M. Grammatikou, and B. Maglariset, Hierarchical anomaly detection in distributed large scale sensor networks, *Proceedings of ISCC*, 2006.
6. F. Martincic and L. Schwiebert, *Distributed event detection in sensor networks*, in *Proceedings of the International Conference on Systems and Networks Communication*, pp. 43–48, 2006.
7. V. Chandola, A. Banerjee, and V. Kumar, *Anomaly Detection: A Survey*, Technical Report, University of Minnesota, 2007.
8. D. Janakiram, A. Mallikarjuna, V. Reddy, and P. Kumar, *Outlier detection in wireless sensor networks using Bayesian belief networks*, *Proceedings of IEEE Comsware*, 2006.
9. W. Wu, X. Cheng, M. Ding, K. Xing, F. Liu, and P. Deng, Localized outlying and boundary data detection in sensor networks, *IEEE Transactions on Knowledge and Data Engineering*, vol. 19, no. 8, pp. 1145–1157, 2007.
10. L.A. Bettencourt, A. Hagberg, and L. Larkey, *Separating the wheat from the chaff: Practical anomaly detection schemes in ecological applications of distributed sensor networks*, *Proceedings of the IEEE International Conference on Distributed Computing in Sensor Systems*, 2007.
11. Y. Hida, P. Huang, and R. Nishtala, *Aggregation Query under Uncertainty in Sensor Networks*, 2003.
12. M. C. Jun, H. Jeong, and C.C.J. Kuo, *Distributed spatio-temporal outlier detection in sensor networks*, in *Proceedings of SPIE*, 2006.
13. B. Sheng, Q. Li, W. Mao, and W. Jin, *Outlier detection in sensor networks*, in *Proceedings of MobiHoc*, 2007.
14. T. Palpanas, D. Papadopoulos, V. Kalogeraki, and D. Gunopulos, Distributed deviation detection in sensor networks, *ACM Special Interest Group on Management of Data*, vol. 32, no. 4, pp. 77–82, 2003.
15. S. Subramaniam, T. Palpanas, D. Papadopoulos, V. Kalogerakiand, and D. Gunopulos, Online outlier detection in sensor data using nonparametric models, *The VLDB Journal – The International Journal of Very Large Data Bases, VLDB*, pp. 187–198, 2006.

16. S. Rajasegarar, C. Leckie, M. Palaniswami, and J.C. Bezdek, *Distributed anomaly detection in wireless sensor networks*, in *Proceedings of IEEE ICCS*, 2006.
17. S. Rajasegarar, C. Leckie, M. Palaniswami, and J.C. Bezdek, *Quarter sphere based distributed anomaly detection in wireless sensor networks*, in *Proceedings of IEEE International Conference on Communications*, pp. 3864–3869, 2007.
18. Y. Chen and S. Li, A lightweight anomaly detection method based on SVDD for wireless sensor network. *Wireless Personal Communications*, vol. 105, pp. 1235–1256, 2019.
19. E. Elnahrawy and B. Nath, *Context-aware sensors*, in *Proceedings of EWSN*, 2004.
20. D. J. Hill, B. S. Minsker, and E. Amir, *Real-time Bayesian anomaly detection for environmental sensor data*, in *Proceedings of 32nd Congress of the International Association of Hydraulic Engineering and Research*, 2007.
21. V. Chatzigiannakis, S. Papavassiliou, M. Grammatikou, and B. Maglariset, *Hierarchical anomaly detection in distributed large scale sensor networks*, in *Proceedings of ISCC*, 2006.
22. R. Chalapathy and S. Chawla, *Deep learning for anomaly detection: A survey*, Online, 2019. Available: https://arxiv.org/abs/1901.03407,2019.
23. D. Kwon, H. Kim, J. Kim, S. C. Suh, I. Kim, and K. J. Kim, A survey of deep learning-based network anomaly detection, *Cluster Computing*, vol. 10, pp. 1–13, 2017.
24. T. Luo and S.G. Nagarajan, *Distributed Anomaly Detection Using Autoencoder Neural networks in WSN for IoT*, IEEE ICC, Kansas City, MO, pp. 1–6, 2018.
25. T. Kieu et al., *Outlier detection for time series with recurrent autoencoder ensembles*, in *Proceedings of the 28th International Joint Conference on Artificial Intelligence (IJCAI)*, pp. 2725–2732, 2019.
26. S. Albelwi and A. Mahmood, A framework for designing the architectures of deep convolutional neural networks, *Entropy*, vol. 19, no. 6, p. 242, 2017.
27. C. Herff and D.J. Krusienski, Extracting features from time series, in Kubben P., Dumontier M., Dekker A. (eds.), *Fundamentals of Clinical Data Science*, Springer, Cham, 2019.
28. Sensor Scope Online. Available: http://sensorscope.epfl.ch/index.php/Mainpage
29. S. Reece, S. Roberts, C. Claxton, and D. Nicholson, *Multi-sensor fault recovery in the presence of known and unknown fault types*, in *12th IEEE International Conference on Information Fusion*, pp. 1695–1703, 2009.
30. F. Shahzad, Pymote 2.0: Development of an interactive python framework for wireless network simulations, *IEEE Internet of Things Journal*, vol. 3, no. 6, pp. 1182–1188, 2016.
31. A. Abid, A. Kachouri, and A. Mahfoudhi, *Anomaly detection through outlier and neighborhood data in wireless sensor networks*, in *Advanced Technologies for Signal and Image Processing (ATSIP), 2nd International Conference*, pp. 26–30, 2016.

9 NEECH

A Novel Energy-Efficient Cluster Head Selecting Protocol in a Wireless Sensor Network

Shelly Bhardwaj and Gurpreet Singh Saini

St. Soldier Institute of Engineering and Technology, Jalandhar, Punjab, India

CONTENTS

9.1 INTRODUCTION

The wireless sensor network (WSN) has facilitated human beings in various sectors of applications [1]. It is characterized by various sensor nodes that are deployed over the area where continuous monitoring of various attributes is done [2]. Figure 9.1a shows the architecture of WSN, and Figure 9.1b demonstrates the components associated with a wireless sensor node. As it can be seen from Figure 9.1a, the sensor nodes communicate to the sink through single-hop or multi-hop communication [3]. A sensor node is a small-sized device with a limited battery stock that runs and performs the functioning of sensing and forwarding the data. The battery of these sensor nodes are irreplaceable in the context that they are mostly used in the areas where humans can't reach [4,5]. Therefore, it is always recommended to use these sensor nodes in the most efficient way that they can be made to run for the longer period [6,7].

DOI: 10.1201/9781003145028-9

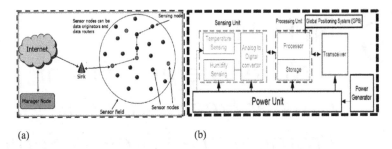

(a)   (b)

FIGURE 9.1   (a) Architecture of WSN (b) Architecture of a sensor node.

Manager node is basically a user which takes a decision based on the threshold defined for various attributes. The primary component out of the essential parts of sensor node is the battery. Once the battery of the sensor node is completely consumed, the node is said to be a dead node. These nodes are deployed for performing in various applications [8]. Among them, the hostile applications and the early detection of some events that may cause a huge damage to the natural resources are the major achievements of the sensor nodes [9]. These applications normally comprise forest fire detection, early detection of flood, and many more [10].The sensor nodes work on the simple operation that they sense the surroundings, and the collected data is being forwarded to the sink where the data is processed for the further required operations that may include the information to the rescue team [11]. The architecture of WSN is discussed in the proceeding subsection that explains how the whole sensor network performs its functioning.

The sensor nodes perform clustering, in which various sensor nodes forms a group which is termed as a cluster. There are two modes of communication in sensor nodes; single hop and multi hop, as shown in Figure 9.2a and b, respectively.

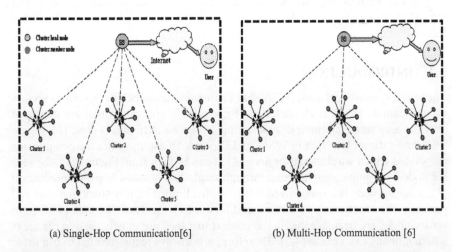

(a) Single-Hop Communication[6]   (b) Multi-Hop Communication [6]

FIGURE 9.2   Modes of communication among sensor nodes: (a) single-hop communication; (b) multi-hop communication [6].

Over the years, various routing protocols have been proposed that aim to enhance the energy efficiency of the network. However, the selection of CH has been the topic of talk in recently proposed work. There are two modes of network for which the various authors have proposed CH, one being homogeneous and other being heterogeneous. The homogeneous network deals with the same configuration, whereas the heterogeneous network has a different configuration of nodes that comprises energy, computing power, sensing range, and various other attributes for the operation.

However, the primary concern is with the Cluster Head (CH) selection. The CH selection depends upon various factors that must be considered [12,13]. However, it is evident that many protocols have considered energy and distance parameters for the CH selection. Whereas, the other parameters, if considered, may improve the overall network efficiency of the network [14].

9.1.1 Major Contributions

The major contributions of the presented work are as follows.

a. A Novel Energy-Efficient Cluster Head (NEECH) selecting protocol is proposed that considers node density factor for the CH selection along with the node's energy and distance factor.
b. The rotation of CH is done when the energy of the selected CH goes below 30% of its total initial energy.
c. The performance of NEECH is compared with the recently proposed protocols.

The rest of the chapter is given as follows. Section 9.2 represents the related work. Section 9.3 gives the proposed work. Section 9.4 discusses the results and discussions. Section 9.5 concludes the chapter.

9.2 RELATED WORK

In this chapter, the research work related to the heterogeneous routing protocols have been discussed. It started from Stable Election Protocol (SEP), the first protocol that reported for the heterogeneous WSN [15]. It selected CH based on weighted function. Thereafter, the trend started for increasing the energy level of various heterogeneous nodes. Distributed Energy Efficient Clustering (DEEC) [16] considered two levels, whereas Energy Efficient Hierarchical Clustering (EEHC) [17] considered three levels of energy heterogeneity. Although various enhancements have been proposed since then, the selection of CH is still a big research problem.

Verma et al. [8] presented MEEC that introduced multiple sinks surrounding the network. While doing so, it tends to enhance the cost of the network. Further, Behera et al. [18] presented a protocol, i.e. R-LEACH, which used residual energy for the selecting CH. It is noted that the better version of traditional LEACH used the random CH selection. The shortcoming of this protocol is found to be non-consideration of distance factor while selecting the head of cluster. Behera et al. [7] proposed i-SEP that improves SEP and its consideration to the evaluating with older protocols makes it not promising.

9.3 WORKING OF NEECH

As discussed, the primary focus of this chapter is on the CH selection. Therefore, the cluster-based routing followed by NEECH is done through the setup phase and steady state phase. The setup phase includes the CH selection, the network formation. The steady state phase includes the data transmission in the network. There are some assumptions which are taken in the network, as follows:

9.3.1 NETWORK ASSUMPTIONS FOR NEECH

a. No entity is supposed to be moved in the network.
b. The nodes are homogeneous.
c. The sink has no limit in the energy resources.
d. The nodes are unknown to the location.
e. The nodes are deployed randomly.
f. The physical factors are not considered.
g. The security aspects of the work are out of the scope of this manuscript.

9.3.2 RADIO ENERGY CONSUMPTION MODEL

The nodes, when communicated in the wireless medium, send data packets wirelessly and transmit l-bit messages. Energy is consumed based on the distance for which data packets are forwarded from one entity to the other. The energy expenditure for transmitting the data packets is given in Equation (9.1):

$$E_{TX}\left(K,d\right) = \begin{cases} KE_{elec} + Kd^2E_{fs} \text{ if } d < d_0 \\ KE_{elec} + Kd^4E_{mp} \text{ if } d \geq d_0 \end{cases} \tag{9.1}$$

Here, E_{elec} represents the energy dissipated per bit for operating the transmitter/ receiver of sensor node. It is noted that when threshold distance (d_0) is greater than distance (d), the free space (E_{fs}) energy model is exploited; otherwise, the multi-path (E_{mp}) energy model is used. The threshold distance (d_0) can be computed as in Equation (9.2):

$$d_0 = \sqrt{\frac{E_{fs}}{E_{mp}}} \tag{9.2}$$

For reception of K-bit data (E_{RX}), the total energy expenditure is measured as in Equation (9.3).

$$E_{RX}\left(K\right) = K \times E_{elec} \tag{9.3}$$

Further, as we know, data aggregation also accounts to the energy expenditure, therefore, the computation of energy spent during aggregating data, is shown as in Equation (9.4).

$$E_{dx}\left(K\right) = p \times K \times E_{da} \tag{9.4}$$

Here, p denotes the packets quantity, and E_{da} denotes the energy exhausted for aggregating the data of one bit.

9.3.3 OPERATION STEPS OF NEECH

The protocol NEECH is operated in a conventional mode similar to the protocols discussed in [13,19]. After the deployment, the nodes are operated in setup and steady-state phase. The deployment scenario, and mathematical model used in these phases are discussed as follow.

 i. *Deployment of sensor nodes.* The sensor nodes in a fixed number is deployed in pre-fixed area where these nodes get connected. These nodes are randomly placed are having same configuration. The total number of nodes that are used are 100 and the area in which the nodes are deployed is 100×100 meter^2.

 ii. *Set-up phase.* In this step, the network formation takes place. The nodes nearer to each other forms a group called a cluster. In this cluster, the selection of CH is done based on the following equations. The LEACH protocol [20] follows Equation (9.5).

$$T(A) = \begin{cases} \dfrac{P}{1-P\left(r \bmod \dfrac{1}{P}\right)} & \text{if } A \in Y \\ 0 & \text{otherwise} \end{cases} \quad (9.5)$$

However, for this chapter, we used the modified equation for the threshold computation for NEECH. The factors, distance, node density and remaining energy are encountered for the selecting CH.

$$Th_{NEECH} = \begin{cases} \dfrac{P_{NEECH}}{1-P_{NEECH}\left(r \bmod \dfrac{1}{P_{NEECH}}\right)} \dfrac{E_{Res} * N_{density}}{D_{(SINK_{NODE})}} & \text{if } N_{NEECH} \in G \\ 0 & \text{otherwise} \end{cases} \quad (9.6)$$

The symbols used in Equation (9.5) are defined as follows. P denotes the optimum probability for the number of CHs, which is defined with value 0.5. The parameter r is the current value of round. Further, in Equation (9.6), the threshold for the NEECH protocol is defined by Th_{NEECH}. The optimum probability or the pre-fixed number of CHs are given by P_{NEECH}. The parameter E_{Res} is the residual energy of the sensor nodes used, while $N_{density}$ is the density of the node which is computed in the similar fashion as computed by protocol discussed in [5]. $D_{(SINK_{NODE})}$ is the distance of a node from the data collecting sink. In Equation (9.6), G represents the group of those nodes which have not become CH. If the threshold value computed in Equation (9.6) is less than the random number, it is selected as CH; else, the node is assigned the role of cluster member.

iii. *Steady State phase*. In this phase, the data transmission is initiated in the network. The cluster member forwards data to the CH, and the CH sends data to the sink. It is noticed that the energy consumed by the CH in data transmitting is more than the cluster member nodes. The reason behind such heavy energy consumption is the fact that CH consumes energy in data aggregating and removing the redundant data. The whole process is shown in Figure 9.3.

The whole operation of NEECH is started with the network formation. Before any phase is commenced, the energy of the node is checked. If the energy value of a node

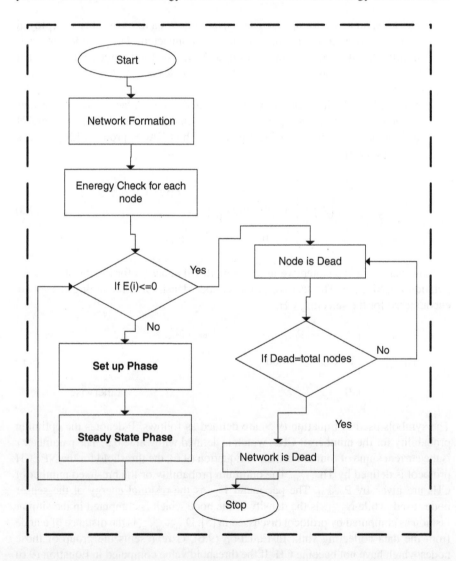

FIGURE 9.3   Flow chart for the proposed work.

TABLE 9.1
Simulation Parameters

Parameter	Value
Network coverage	(100, 100)m
BS location	(50, 50)m
Node Number	100
Initial energy (Quantity)	In Joules 0.5
E_{elc}	50 nJ/bit
E_{efs}	10 pJ/bit/m^2
E_{mp}	0.0013 pJ/bit/m^4
d_0	87 m
E_{da}	5 nJ/bit/signal
Data packet size	4000 bits

is more than the threshold value, only then is it considered for the CH selection. If the total energy of all nodes is dead, the network stops functioning.

9.4   RESULTS AND DISCUSSION

There are some performance metrics on which the result section of the proposed protocol is discussed. The simulation parameters are discussed in Table 9.1.

9.4.1   PERFORMANCE METRICS

Various performance metrics are considered while comparing the performance of the proposed protocol with the other protocols.

a. *Stability period.* The number of rounds completed when first node is dead, i.e., the energy of any node at the first place, becomes zero after going through the number of rounds. This parameter helps in determining the network performance as the longer the stability period, the more that reliability of the network is assured. Therefore, various routing algorithms have as their main focus to enhance stability period to the maximum extent. In Figure 9.4, the stability period of NEECH is more than that of the RES-LEACH, I-SEP and MEEC protocols. The graph of alive node versus rounds and dead nodes versus rounds is shown in Figures 9.4 and 9.5, respectively. It is evident that NEECH has outperformed other protocols in both metrics (Figures 9.6 and 9.7).

b. *Half Dead Node.* This is the number of rounds completed till half of the nodes are dead in the network. The number of rounds covered by these nodes determines the efficiency of the network. Figures 9.4 and 9.5 show the protocol NEECH has more efficiency as compared to other protocols.

c. *Network lifetime.* This is the number of rounds covered until last node is dead or the whole number of nodes are dead in the network. The network lifetime is a very essential parameter to determine the performance of any protocol, as it

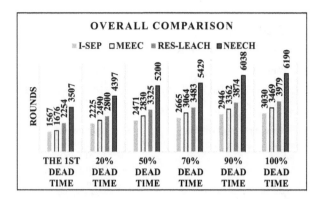

FIGURE 9.4   The overall performance of NEECH.

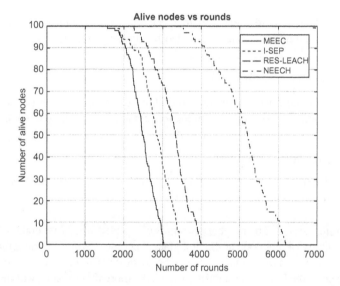

FIGURE 9.5   Alive nodes vs. rounds.

encourages the various applications for which the sensor network could be designed. Figure 9.4 shows the overall performance of NEECH as compared to other protocols.

d. *Network remaining energy.* The protocol NEECH has also performed very well for covering a greater number of rounds as compared to protocols discussed previously. The energy of the nodes is preserved for more number of rounds.

e. *Throughput or number of packets sent to the sink.* As the number of rounds are proceeded, the nodes start sending their data packets to the sink. Therefore, it

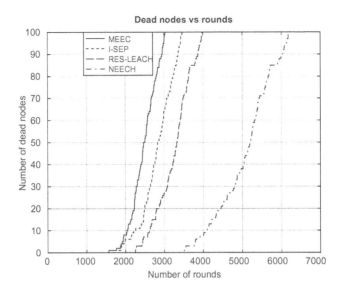

FIGURE 9.6   Dead nodes vs. rounds.

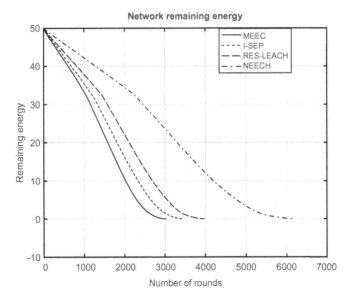

FIGURE 9.7   Network remaining energy.

can be defined as the number of data packets sent to the sink. This parameter shows the reliability for data collection in the network provided by the given technique. Figure 9.8 shows that the NEECH protocol outperforms in delivering more packets to the sink.

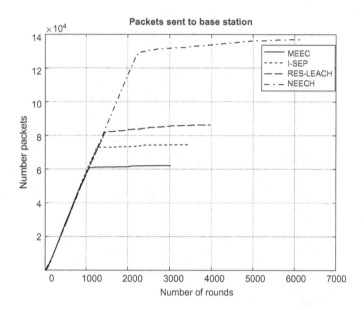

FIGURE 9.8 Throughput.

9.5 CONCLUSION

Since the development of WSN, the nodes have been facing the concern of limited battery. Once the nodes are equipped with the battery, they start consuming energy with the passage of rounds. Hence, in this chapter, we proposed the Novel Energy Efficient CH selecting (NEECH) protocol that selects CH based on different parameters, namely, distance, energy stock, and density of nodes. We have given a mathematical model and also explained the radio energy model. The results are presented in a comprehensive manner, and every metric has been explained. Finally, it is observed that the proposed protocol helps in the improving the network performance drastically. In future, we will focus on the sink mobility scenario in the WSN that will help in attaining better performance of the whole network.

REFERENCES

1. S. Verma, N. Sood, and A. K. Sharma, "Cost-effective cluster-based energy efficient routing for green wireless sensor network," *Recent Advances in Computer Science and Communications*, vol. 14, no. 4, pp. 1040–1050, 2021.
2. S. R. Pokhrel, S. Verma, S. Garg, A. K. Sharma, and J. Choi, "An efficient clustering framework for massive sensor networking in industrial IoT," *IEEE Transactions on Industrial Informatics*, 2020, doi:10.1109/tii.2020.3006276.
3. B. M. Sahoo, T. Amgoth, and H. M. Pandey, "Particle swarm optimization based energy efficient clustering and sink mobility in heterogeneous wireless sensor network," *Ad Hoc Networks*, vol. 106, p. 102237, 2020.
4. S. Verma, S. Kaur, A. K. Sharma, A. Kathuria, and M. J. Piran, "Dual sink-based optimized sensing for intelligent transportation systems," *IEEE Sensors Journal*, 2020. doi:10.1109/JSEN.2020.3012478.

5. B. M. Sahoo, H. M. Pandey, and T. Amgoth. "GAPSO-H: A hybrid approach towards optimizing the cluster based routing in wireless sensor network." *Swarm and Evolutionary Computation*, vol. 60, pp. 100772, 2020.

6. S. Verma, N. Sood, and A. K. Sharma, "A novelistic approach for energy efficient routing using single and multiple data sinks in heterogeneous wireless sensor network," *Peer-to-Peer Networking and Applications*, vol. 12, no. 5, pp. 1110–1136, 2019.

7. T. M. Behera, S. K. Mohapatra, U. C. Samal, M. S. Khan, M. Daneshmand, and A. H. Gandomi, "I-sep: An improved routing protocol for heterogeneous WSN for IoT-based environmental monitoring," *IEEE Internet of Things Journal*, vol. 7, no. 1, pp. 710–717, 2019.

8. S. Verma, N. Sood, and A. K. Sharma, "Design of a novel routing architecture for harsh environment monitoring in heterogeneous WSN," *IET Wireless Sensor Systems*, vol. 8, no. 6, pp. 284–294, 2018.

9. S. Verma, R. Mehta, D. Sharma, and K. Sharma, "Wireless sensor network and hierarchical routing protocols: A review," *International Journal of Computer Trends and Technology (IJCTT)*, vol. 4, no. 8, pp. 2411–2416, 2013.

10. D. Pant, S. Verma, and P. Dhuliya, *"A study on disaster detection and management using WSN in Himalayan region of Uttarakhand,"* in *2017 3rd International Conference on Advances in Computing, Communication & Automation (ICACCA) (Fall)*, Dehradun, India, 2017, pp. 1–6.

11. A. Z. Abbasi, N. Islam, and Z. A. Shaikh, "A review of wireless sensors and networks' applications in agriculture," *Computer Standards and Interfaces*, vol. 36, no. 2, pp. 263–270, 2014.

12. S. Verma and K. Sharma, "Energy efficient zone divided and energy balanced clustering routing protocol (EEZECR) in wireless sensor network," *Circuits and Systems: An International Journal (CSIJ)*, vol. 1, no. 1, 2014.

13. T. M. Behera, U. C. Samal, and S. K. Mohapatra, "Energy-efficient modified LEACH protocol for IoT application," *IET Wireless Sensor Systems*, vol. 8, no. 5, pp. 223–228, 2018.

14. S. Verma, N. Sood, and A. K. Sharma, "QoS provisioning-based routing protocols using multiple data sink in IoT-based WSN," *Modern Physics Letters A*, vol. 34, no. 29, p. 1950235, 2019.

15. B. M. Sahoo, R. K. Rout, S. Umer, and H. M. Pandey. *"ANT colony optimization based optimal path selection and data gathering in WSN,"* in *2020 International Conference on Computation, Automation and Knowledge Management (ICCAKM)*, pp. 113–119. IEEE, 2020.

16. L. Qing, Q. Zhu, and M. Wang, "Design of a distributed energy-efficient clustering algorithm for heterogeneous wireless sensor networks," *Computer Communications*, vol. 29, no. 12, pp. 2230–2237, 2006.

17. D. Kumar, T. C. Aseri, and R. B. Patel, "EEHC: Energy efficient heterogeneous clustered scheme for wireless sensor networks," *Computer Communications*, vol. 32, no. 4, pp. 662–667, 2009.

18. T. M. Behera, S. K. Mohapatra, U. C. Samal, M. S. Khan, M. Daneshmand, and A. H. Gandomi, "Residual energy-based cluster-head selection in WSNs for IoT application," *IEEE Internet of Things Journal*, vol. 6, no. 3, pp. 5132–5139, 2019.

19. S. Verma, S. Kaur, M. A. Khan, and P. S. Sehdev, "Towards green communication in 6G-enabled massive Internet of Things," *IEEE Internet of Things Journal*, vol. 8, no. 7, pp. 5408–5415, 2020, doi:10.1109/JIOT.2020.3038804.

20. W. R. Heinzelman, A. Chandrakasan, and H. Balakrishnan, *"Energy-efficient communication protocol for wireless microsensor networks,"* in *Proceedings of the 33rd Annual Hawaii International Conference on System Sciences, 2000*, 2000, p. 10, Accessed: September 18, 2017. [Online]. Available: http://ieeexplore.ieee.org/abstract/document/926982/.

10 An Efficient Model for Toxic Gas Detection and Monitoring Using Cloud and Sensor Network

A. M. Senthil Kumar, Yamini Pemmasani, Haritha Venkata Naga Siva Sruthi Addanki, Vasantha Sravani, and Dama Srinu

Koneru Lakshmaiah Education Foundation, Guntur, Andhra Pradesh, India

CONTENTS

10.1 INTRODUCTION

Gas has become available to the public in many forms. In the modern world, most households tend to store several LPG gas canisters which are used for cooking. Even if the physical presence of canisters is not present within a house, the house itself will have LPG pipes to support the kitchen stoves or appliances. Due to several reasons, the use of gas has drastically increased over the past few years. Although the simplicity to access gas for day-to-day purposes has become an advantage, many catastrophes have taken place in the world due to simple gas leaks. Many have seen incidents in which thousands of people have lost their lives due to the inflammation of leaked gas. The after-effects are severe in nature, and it takes a considerably large force to

DOI: 10.1201/9781003145028-10

maintain such events to prevent casualties. This is just a simple example of a gas leak in the public world.

Other major leaks take place in mines or deep underground workplaces which are prone to poisonous gases. Several deaths in mines have been caused by the lack of gas detection. Underground mines tend to have pockets of dangerously poisoned gases which rupture and leak into mines. Although miners are well prepared for these scenarios with proper equipment, they are unable to analyze the situation quickly enough to prevent any casualties. By the time that the miners are able to verify that the mine is filled with poisonous or dangerous gas, several people will have died; this has occurred several times in the past. Although the use of gas detectors has been able to prevent these cases in several scenarios, these detectors still lack the ability to alert the proper authorities when they occur. This is why it's time to create a system which can constantly store all of its data on a cloud platform so that it can be constantly monitored by people from anywhere in the world.

In this chapter, we will rely upon ThingSpeak to store all of the data. ThingSpeak is an online cloud platform which is able to extract data from sensors and store all of its data upon its cloud servers. Another advantage of using this platform is its built-in ability to constantly analyze all of the data being stored to create visual charts or graphs for the user to understand. This is important when one deals with gas leaks, as it will be able to constantly monitor the given area and visually understand the change in gas levels over a given span of time. It will be able to easily see changes within a specific type of gas with a sudden rapid spike in the graphs generated.

The main advantage of this system is its ability to use the latest cloud technology. A cloud allows us to store the data in a virtual storage space which is located remotely in a server. The modern gas sensors are able only to detect the increase of a specific type of gas to warn the people in its residence. However, this will not be able to alert officials outside, and often in cases, they may be trapped inside the area without access to the outside world. Due to this, it will often be able only to understand that an issue has occurred after many calamities have taken place. The implementation of this cloud storage allows us not only to constantly view and analyze the data, but to use it in the future too.

Suppose that one could create an algorithm to identify intervals in which a specific type of gas was to be spiked in an area that could use all the previous data stored in the cloud to accomplish this. This is impossible merely with sensors installed on the premises as they are small in size factor and don't carry large storage devices within them to record the data. Even on the occasion that they do, the data will be constantly wiped and rewritten so that one can look back in a small time frame. This model will also have various other sensors which are able to sense other aspects of nature, such as temperature or humidity, to gain an understanding of the environment as well. Although they may seem ineffective in the leakage of gas, one can look further into these statistics to analyze how they may affect one another.

This model will also have a built-in threshold limit which will automatically intimate the concerned individuals in case a gas leak is identified so that proper actions are taken immediately. To accomplish this, one will be using various applications and hardware components.

Now that we have a clear understanding of the problem at hand and how to approach it, we will now continue onto the procedure outlined in this chapter.

Through this chapter, one will be able to look into the modern works in this domain with the literature review, which has been neatly written down in Section 10.2. From there, it will start to explain the model that has been proposed to create in Section 10.3, followed by the algorithm in Section 10.4. Once if it gives a clear idea on how it been decided to approach this issue it will run us through a procedure through how it is created about this prototype in Section 10.5 which will be followed by the various results that have obtained in Section 10.6. Conclusion and future work is discussed in Section 10.7.

10.2 LITERATURE REVIEW

The use of various alarm systems in the detection of hazardous gases has become common over the past decade. Although these devices which act as the alarm systems are not able to completely erase the casualties in the event of any gas leakage, they have been able to alert people in the vicinity.

The various alarm systems which have been made so far are all unique and independent from one another. Their internal architecture, as well as the method of intimation or data storage, largely differs not only with this proposed model but with those that exist as well. While some of the researchers have decided to build their prototype with the help of an Arduino [1,2], others have decided to use a Raspberry Pi [2–4] board similar to what this model used.

To place aside the core component of this prototype, one can observe that all of the sensors used throughout this chapter have all been constant. The MQ2 has been found in almost of the currently proposed models. However, there have been models created with more gas detection sensors, such as the MQ7 [5] and the MQ135 [6], in order to get a better understanding of their surroundings. Another sensor used in such cases was a radiation sensor [7,8] which was added to the prototypes having industrial areas or mines in mind so that their prototype could react to radioactive situations as well [9].

Across all of the currently established prototypes, it has been able to establish the fact that all of their data is being stored within some sort of database physically or virtually [10] with the help of the internet. Among the cases in which the prototypes are placed in mines deep [11] under the surface with no signal, they have implemented a physical storage system [12] and wiring to allow them to store data onboard and also transmit it when required using the cables. However, they all lack in terms of constant monitoring of an are without physical presence [13].

The major advantage to the prototype that has been proposed is its ability to constantly monitor an environment without physically having to be there. Among the limitations of the various models that were looked into was the fact that an alarm system was used only to alert the people in the vicinity, not those outside. In the scenario that these alarms were to be triggered, the people in the area would be alerted, but the outside world would not know about the scenario until someone from the inside called or notified people who were not there. Due to this, it is not able to get the proper help or support required until the proper officials are called.

To overcome this issue, the data will be constantly stored from the sensor and updated on an online cloud platform which anyone physically anywhere may access by using this application. A simple Android application will also be created with the help of an MIT app inventor which is able to retrieve the data stored in the cloud platform and displays all of it neatly to the user. Another advantage is its ability to send WhatsApp messages to the required individuals in case a gas leakage took place.

All of these implementations may seem small but the time they will save may prevent many casualties in the leakage of any form of gas. Cloud platforms can be used in various applications for efficient results [14–21]. This cloud platform is able to help us constantly store and analyze the data, while the application is able to help us retrieve and view the entire statistic. This built-in alarm system won't require people to contact the higher officials, as they will automatically be notified via a WhatsApp message.

10.3   PROPOSED GAS DETECTION AND MONITORING MODEL

In this section, one will use several different components along with hardware. However, the heart of this chapter will be the Raspberry Pi board in which one will be executing all of the programs. The gases in the environment are detected with the help of an MQ2 sensor, while on the other side, the humidity and tempera-ture values are constantly taken using a DHT11 module. The Raspberry Pi will be communicating with the various different applications such as the app that will be created along with the ThingSpeak cloud which will be used to store all of the data. Apart from these, the main module will also be working with a Cisco PL application along with a Twillio API to take care of the alerts and other notifications.

To get a clear picture of how all of the components and hardware are related, take a look at Figure 10.1, which illustrate the entire idea in the chapter.

- Components: Raspberry Pi, Sensors for detection, ADC
- OS: IOT PL-App Image
- APIs: Twilio
- Cloud: ThingSpeak
- Technologies: Python, PL-App Launcher

10.4   PROPOSED GAS DETECTION ALGORITHM

Now that we have a clear understanding of the internal structure of the hardware along with the software within this model, let us now look at the pseudo-code for the algorithm that one will implement in this model.

Step 1: Read Data from Sensors

The first step in this algorithm is making sure that one has properly connected the modules to the proper input pins of the Raspberry Pi boards to make sure the pins can retrieve the data from them. One will be calling these pins within this algorithm to retrieve the data from them.

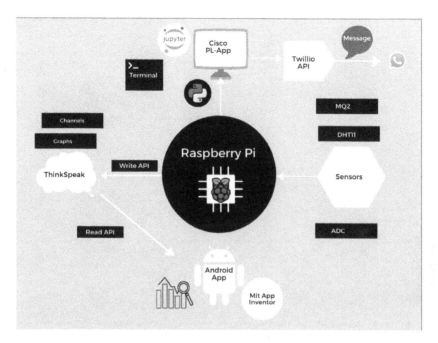

FIGURE 10.1 Proposed model.

Step 2: Convert Analog data received from the sensor to Digital data using ADC

The data that will be received from the modules will be in the form of an analog signal. To make sure that this algorithm is able to use this data, one will have to use an ADC to convert the data into a digital format to be used within this algorithm.

Step 3: Send Digital data from ADC to Raspberry Pi

Once the data has been converted into a digital format, it's now time to send the data to the Raspberry Pi so that it can be further worked within this algorithm. It's now time to make sure that the data received is not incorrect or corrupted before moving on to further use of the data.

Step 4: Sending data from Raspberry Pi to ThingSpeak in Cisco PL-App using Python

Once the data has been read by the Raspberry Pi and is retrieved, one will now start to write code using the API keys of the sensors along with the ThingSpeak account so that the data is collected and stored on the cloud. It will transfer the data with the help of the Cisco PL-App.

Step 5: Read data from ThingSpeak using Read API and View in Android App

Once the data is stored in the ThingSpeak cloud, it's time to now move onto the next step of retrieving it for this application. In this application, one will be using the Read API keys from ThingSpeak to look at all of the data and the graphs which have been constantly stored and updated in the cloud platform

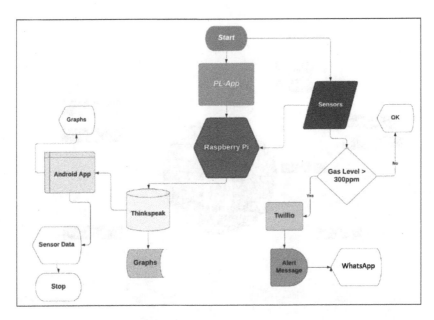

FIGURE 10.2   Proposed algorithm flowchart.

Step 6: Send alert messages when gas levels are high with Twillio API using Python

The final step in this process is to activate the emergency message alerts. One will be implementing these emergency messages using the Twillio API. These emergency messages will be sent to the WhatsApp number that has been specified within this program. This number can be updated within the python code and it is even able to allow more than one person to receive it by adding extra commands. In Figure 10.2, the proposed algorithm flowchart is shown. It will be able to get a closer understanding of this algorithm along with the various conditions which may take effect on the outputs as well.

10.5   IMPLEMENTATION

There is a total of five steps that one must first follow in order to create an efficient model that is able to not only detect and alert us during the presence of a gas leak but also store all of the data accumulated from the sensors on a cloud platform and have it analyzed. The various steps in this section will deal with booting the Raspberry Pi and installing all the proper applications all the way to creating this application to view the data. Let's take a look into it one step at a time.

10.5.1   BOOTING THE RASPBERRY PI

The first step is to properly install the image file in the Raspberry Pi board and make sure that it is properly booted up. In this section, one will be installing an image file known as Cisco PL App Launcher in the Raspberry Pi device, as this will allow us to look into all the application devices as an API. This is very crucial for this chapter, as it will be using these API keys to link the sensors to the cloud and vice versa. Once it has properly loaded the bootable image into the Raspberry Pi and made sure that the operating system is properly functioning and booting up with all the files, it's time to proceed to the next step.

10.5.2 SECURING ALL HARDWARE CONNECTIONS

Since the Raspberry Pi board is now able to boot up and access all of the applications that require, it's time to make all the required hardware connections. The circuit diagram is shown in Figure 10.3. In this model, it's been decided to use various sensors to constantly monitor the environment but it would be likely to specifically highlight two of them. One will be using the MQ2 sensor to monitor the gas present in the room, while a DHT11 sensor will constantly monitor the humidity as well as the temperature of the environment. To get a better understanding of how all the components should be connected together, take a look at the circuit diagram in Figure 10.3.

Once all the connections are properly established to the proper pins on to the Raspberry Pi board, we can now code the board to input all the data from the sensors. One will be doing this using the built-in Python compiler on the board and run various commands to import all the required packages for the sensors as well as programs. Once the Raspberry Pi board is able to constantly retrieve data from the environment using the sensors with no issues and a high accuracy we can then move onto the next crucial part of this process.

10.5.3 IMPORTING SENSOR DATA ONTO THE CLOUD PLATFORM

In this section, the platform known as ThingSpeak is used. This platform enables us to connect the sensors to an online cloud database and automatically store the data in it. This process is done with the use of various API keys. One will be creating a Python code which will store the API keys of the sensors and have them route all of their data to the ThingSpeak cloud using the Wi-Fi connection that the Raspberry Pi board has connected to.

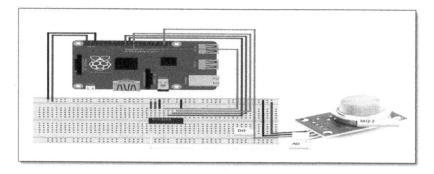

FIGURE 10.3 Circuit diagram.

It will be able to check whether the data has been sent to the cloud by logging into the ThingSpeak cloud account and looking into all of the data which is available on the cloud. Once the sensors have been properly integrated the outputs of all the sensors, it will be visible on ThingSpeak, and one will also be able to see the statistical data in the form of a line graph over a given span of time.

10.5.4 ENABLING TWILLIO

The final step to complete this model is to set the threshold values to react to. One will be accomplished with the use of software known as Twillio. This software allows us to fix in the threshold values for all of the sensors and makes sure that in the occasion in which these thresholds are passed, a message or notification is sent to the concerned individuals.

10.5.5 CREATING AN APPLICATION

Although one can view all of the data from the sensors, this will be inconvenient as this data will be assessable only to the person with the ThingSpeak account. To overcome this problem, an application will be created which will allow us to view all of the data. To create an application, it is decided to use the MIT app inventor, which has a simple approach to building applications in which one will be able to view all of the data that has been stored in the cloud.

This application can be downloaded by anyone, and with the proper authority, they will be able to access all of the data. One will also be able to look at all of the various graphs that have been drawn using the statistical information from the sensors to get a visual idea on the gas rates in a specific area or environment. Taking all of these into consideration, one can then create an application that can be used universally without the need of having to log into ThingSpeak.

10.6 RESULTS

The result of the proposed gas detection system is evaluated in terms of temperature, humidity, and gas level. The results for this model have all been shown in the various figures available in this section. Figures 10.4–10.6 are all the graphs obtained from

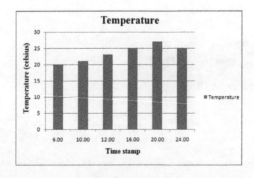

FIGURE 10.4 Temperature level reading from sensors.

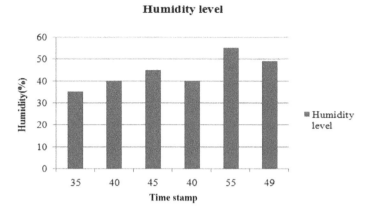

FIGURE 10.5 Humidity level reading from sensors.

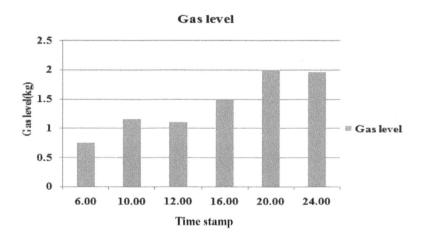

FIGURE 10.6 Gas level reading from sensors.

the reading from the sensor readings. These graphs have been automatically created by ThingSpeak so that the user is able to get a clear understanding of all the data that is being stored within the cloud platform. These graphs will constantly update themselves with the data that is being sent to the cloud storage.

Figure 10.4 shows the temperature data which is read from the sensors present in the proposed model. Temperature data is shown in the Y-axis and date is shown in the X-axis. The values are presented in the graph based on date wise.

The humidity level readings from sensors are shown in Figure 10.5. In the X-axis, the date is shown, and in the Y-axis, the humidity level is shown. The humidity data of the proposed model is shown in the graph date wise.

Figure 10.6 shows the gas level sensor data which is read from the proposed model. Sensor gas level data is shown in the Y-axis, and date information is shown in the X-axis. The gas level values are presented in the graph date wise.

FIGURE 10.7   Application to display sensor reading.

As has been previously mentioned regarding the use of an application within this chapter one can observe the user interface and all of the details that it will show the user in Figure 10.7. This application will be constantly updated, and the reading will be checked at any time with the use of the application. If the level of the gas in the locality is to suddenly raise higher than the threshold value, there will be an automated warning message sent to the concerned authorities' WhatsApp number as shown in Figure 10.8.

FIGURE 10.8   Alert on suspicious gas levels.

10.7 CONCLUSION

Through this chapter, one will be working with various sensors that are involved with detecting the presence of various gases in an environment. Although the modern world has simply overlooked the issues of gas leaks with the constant calamities that have been taking place due to these, it is time for us to take precautions as well. Using this model, we are able to properly monitor the given environment from anywhere with the help of the ThingSpeak cloud. Many modern models are confined to being able to store data for a long term and have to constantly clear their memory to overcome this issue. However, it is believed that if more systems started to take advantage of these cloud platforms, they will be able to store as much as data that is required so that if they are required to in the future, the data can be easily retrieved without having to search for lost data. It's a strong belief that the usage of these cloud storages and platforms is endless and will soon start playing a major role in the world around us.

As future work, more gases can be included in the model for detection. Also, more output parameters can be considered in the gas detection model. While machine learning is not a new technique, interest in the field has exploded in recent years. This resurgence comes on the back of a series of breakthroughs, with deep learning setting new records for accuracy in areas such as speech and language recognition, and computer vision. Two factors primarily make these successes possible. One is the vast quantities of images, speech, video, and text that is accessible to researchers looking to train machine-learning systems. Today, anyone with an internet connection can use these clusters to train machine learning models via cloud services. As the use of machine learning has taken off, so companies are now creating specialized hardware tailored to running and training machine-learning models. As hardware becomes increasingly specialized and machine-learning software frameworks are refined, it's becoming increasingly common for ML tasks to be carried out on consumer-grade phones and computers, rather than in cloud data centers.

REFERENCES

1. Rathod, S. B., Reddy, V., Krishna, N., Dynamic Framework for Secure VM Migration Over Cloud Computing. *Journal of Information Processing Systems*, Vol. 13, No. 3, pp. 476–490, Jun. 2017.
2. Perez-Botero, D., A Brief Tutorial on Live Virtual Machine Migration From a Security Perspective, Princeton University, Princeton, NJ, 2011. Published under licence by IOP Publishing Ltd.IOP Conference Series: Materials Science and Engineering, Volume 981, International Conference on Recent Advancements in Engineering and Management (ICRAEM-2020) 9–10 October 2020, Warangal, India.
3. Babukarthik, R. G., Kumar, J. S., Amudhavel, J., Secure Data Storage and Sharing In Cloud: VM Scheduling, *IIOAB Journal*, 2018, published under license by IOP Publishing Ltd. IOP Conference Series: Materials Science and Engineering, Volume 981, International Conference on Recent Advancements in Engineering and Management (ICRAEM-2020) October 9–10, 2020, Warangal, India.
4. Praveen, S. P., Rao, K. T., Janakiramaiah, B. Effective Allocation of Resources and Task Scheduling in Cloud Environment Using Social Group Optimization. *Arabian Journal for Science and Engineering; Research Article—Special Issue—Computer Engineering and Computer Science*, Vol. 43, No. 8, pp. 4265–4272, 2018.

5. Kalaipriyan, T., Amudhavel, J., Pothula, S. Solving Virtual Machine Placement in Cloud Data Centre Based on Novel Firefly Algorithm. *Bioscience Biotechnology Research Communications*, Vol. 11, No. 1, pp. 48–53, (2018).

6. Bashir, A. K., Arul, R., Basheer, S., Raja, G., Jayaraman, R., Qureshi, N. M. F. An Optimal Multitier Resource Allocation of Cloud Ran in 5G Using Machine Learning. 2018. ISSN 2161-3915.

7. Asvany, T., Amudhavel, J., Pothula, S. Shuffled Frog-Leaping Algorithm SFLA for Solving Load Balancing Problem Through Fog Computing in Cloud Servers. *Bioscience Biotechnology Research Communications*, Vol. 1, Special Issue No. 1, pp. 44–47, 2018.

8. Balakrishna, G., Moparthi, N. R. ESBL: Design and Implement a Cloud Integrated Framework for IoT Load Balancing. *International Journal of Computers Communications & Control*, Vol. 14, No. 4, pp. 459–474, August 2019. ISSN 1841-9836, e-ISSN 1841-9844.

9. Lavanya, K., Reddy, L. S. S., Eswara Reddy, B. Distributed Based Serial Regression Multiple Imputation for High Dimensional Multivariate Data in Multicore Environment of Cloud. *International Journal of Ambient Computing and Intelligence (IJACI)*, Vol. 10, No. 2, p. 17, 2019. doi:10.4018/IJACI.2019040105.

10. Rao, P. R., Sucharita, V. A Framework to Automate Cloud Based Service Attacks Detection and Prevention. *International Journal of Advanced Computer Science and Applications (IJACSA)*, Vol. 10, No. 2, 2019.

11. Lavanya, K., Reddy, L. S. S., Reddy, B. E. Distributed Based Serial Regression Multiple Imputation for High Dimensional Multivariate Data in Multicore Environment of Cloud. *International Journal of Ambient Computing and Intelligence (IJACI)*, Vol. 10, No. 2, pp. 1–17, April–June 2019.

12. Sharmila, P., Danapaquiame, N., Subhapriya, R., Janakiram, A., Amudhavel, J. Secure Data Process in Distributed Cloud Computing. *Bioscience Biotechnology Research Communications*, Vol. 11, No. 1, pp. 75–84, 2018.

13. Gavvala, S. K., Jatoth, C., Gangadharan, G. R., Buyya, R. QoS-Aware Cloud Service Composition Using Eagle Strategy. *Future Generation Computer Systems—The International Journal of Escience*, Vol. 90, pp. 273–290, 2019.

14. Senthil Kumar, A. M., Venkatesan, M. Multi-Objective Task Scheduling Using Hybrid Genetic-Ant Colony Optimization Algorithm in Cloud Environment. *Wireless Personal Communications*, Vol. 107, No. 4, pp. 1835–1848, 2019.

15. Senthil Kumar, A. M., Venkatesan, M. Task Scheduling in a Cloud Computing Environment Using HGPSO Algorithm. *Cluster Computing*, Vol. 22, No. Suppl 1, pp. 2179–2185, 2019.

16. Senthil Kumar, A. M., Venkatesan, M. A Novel Based Resource Allocation Method on Cloud Computing Environment Using Hybrid Differential Evolution Algorithm. *Journal of Computational and Theoretical Nanoscience*, Vol. 14, No. 11, pp. 5322–5326, 2017.

17. Buyya, R., Yeo, C. S., Venugopal, S., Broberg, J., Brandic I. Cloud Computing and Emerging IT Platforms: Vision, Hype, and Reality for Delivering Computing as the 5th Utility. *Future Generation Computer Systems*, Vol. 25, No. 6, pp. 599–616, 2009.

18. Jian, C., Chen, J., Ping, J., Zhang, M. An Improved Chaotic Bat Swarm Scheduling Learning Model on Edge Computing. *IEEE Access*, Vol. 7, pp. 58602–58610, 2019.

19. Shi, Y., Luo, L., Guang, H. *Research on Scheduling of Cloud Manufacturing Resources Based on Bat Algorithm and Cellular Automata. IEEE International Conference on Smart Manufacturing, Industrial & Logistics Engineering (SMILE)*, pp. 174–177, 2019.

20. Pappula, L., Ghosh, D. Cat Swarm Optimization with Normal Mutation for Fast Convergence of Multimodal Functions. *Applied Soft Computing*, Vol. 66, pp. 473–491, 2018.

21. Reshma, T., Reddy, K. V., Pratap, D., Agilan, V. Parameters Optimization Using Fuzzy Rule Based Multi-Objective Genetic Algorithm for an Event Based Rainfall-Runoff Model. *Water Resources Management: An International Journal, Published for the European Water Resources Association (EWRA)*, Vol. 32, No. 4, pp. 1501–1516, 2018.

11 Particle Swarm Intelligence-Based Localization Algorithms in Wireless Sensor Networks

Ravichander Janapati

SR University, Warangal, Telangana, India

Ravi Kumar Jatoth

National Institute of Technology, Warangal, Telangana, India

A. Brahmananda Reddy

VNR VJIET College, Hyderabad, Telangana, India

Ch. Balaswamy

Gudlavalleru Engineering College, Gudlavalleru, Andhra Pradesh, India

CONTENTS

DOI: 10.1201/9781003145028-11

11.1 INTRODUCTION

Recent advancements in wireless networking have gained greater prominence in the field of wireless sensor network (WSN), which carries out sensing, computing, and data transmission from the source node to the sink node [1,2]. WSN is a technology that has driven a social revolution in people's everyday lives. The WSN is used to track farming, military, disaster management, and hospitals in various applications [3]. In the WSN, nodes communicate and send data through the central node to the actual site. WSN improves service coverage, fault detection capability, reliability and data transmission. In WSN nodes the position data of certain nodes cannot be called agent nodes, while known position information nodes are called anchor nodes, and they are deployed by random means. A location is known for finding the exact position of the agent node [4–6].

In health monitoring applications, goal tracking, and transferring data from source to destination, the position of a node is significant. The Global Positioning System (GPS) can be used to locate the nodes, but it can operate well only in an indoor environment, due to many factors such as accuracy and costs. Indoor location systems provide a wide range of protection, indoor navigation, emergency, localization reports, sports, hospitals, and mobile nodes monitoring applications. Indoors, the signal strength can be degraded due to noise and multi-way effects which lead to inaccurate estimates. Accurate indoor environment position measurements are, therefore, an important activity. To find the exact node location, localization algorithms have been used. Different routers are not available for WSN nodes from long distances, unlike in fixed networks [7,8].

Each sensor node acts as a route from source to destination to transmit the data. Routing algorithms have been developed to transmit information from source to destination. These are classified efficiently in three forms:

1. Central data routing
2. Routing hierarchy network
3. Location position-based routing

Each node behaves in the same way in data-centric routing and works in cooperation to carry out the mission. In this routing node, information from all other nodes is obtained and data transmitted. In a hierarchical routing, the data are sensed with minimal energy nodes and the maximum energy required transmitting data. The layered concept makes this routing. Selected layer cluster heads and routing by another layer are introduced. Simple locations and accurate, energy-efficient, and scalable data are transmitted via localized routing algorithms [6].

Conventional routing algorithms, in comparison, are more complicated, require more bandwidth and resources, and are not scalable. The accuracy of positioning of location-based routing algorithms plays a major role. Inaccurate node positions for application routing would contribute to network efficiency degradation in the form of PDR, throughput and life span, etc. [9].

This chapter includes the following significant contributions:

- Knowing the same node position indoors
- Optimal node reference range for efficient position

- Implementation of Particle swarm based localization algorithms, CDPSO locally operating routing and particle swarm optimization helped adaptive extended Kalman filter (PSO-AKF), and enhanced localized WSN Routing for enhanced location-based routing and monitoring with PSO assisted AKF (ELR).

11.1.1 OBJECTIVES OF THE CHAPTER

The key objectives of the research are as follows:

1. Development of the distributed localization algorithm based on PSO.
2. Range of sensor nodes that provide accurate position.
3. To build a methodology that can identify many unknown nodes with greater accuracy, less error, less costly computationally, and decreased energy usage and overhead.
4. Selection of route using accurate location.
5. Routing of data from source to destination with improved PDR, throughput, and energy efficiency.

11.1.2 SCOPE OF THE CHAPTER

The aim of this investigation is to develop a localization algorithm and location based routing algorithms [10–15]. The following research has been carried out to achieve the goal:

1. CDPSO with CRB algorithm is proposed for finding node location.
2. The accuracy of location has been enhanced by PSO-AKF with optimum references.
3. Routing the data from source to destination uses CDPSO-CRB.
4. Enhanced localized routing algorithm has been proposed using PSO-AKF.

The flow of the research work is shown in Figure 11.1.

11.2 EXISTING LOCALIZATION ALGORITHMS

Several of the existing methods will be explained in this section. In the outdoor world, the Global Positioning System (GPS) is used for WSN localization; however, it does not have reliable results and uses more power in the indoor environment [16]. The multiliterate method is used to estimate node location. In this approach, nodes use all available nodes as reference nodes. Hence, location error is greater. The nearest three reference algorithms use three minimum measured distance nodes as reference nodes [17]. A comparison of existing localization algorithms is given in Table 11.1.

11.3 COOPERATIVE DISTRIBUTIVE PARTICLE SWARM OPTIMIZATION (CDPSO)

PSO is an algorithm of swarm intelligence, the social actions of birds, and the fish school [18]. It produces a series of solutions known as particles. Since the PSO

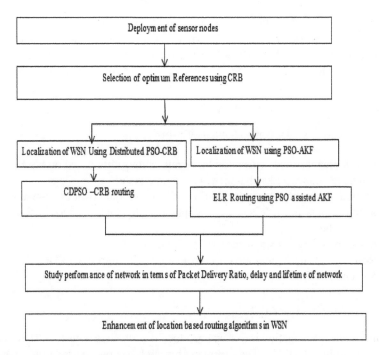

FIGURE 11.1   Block diagram of proposed research work.

TABLE 11.1
Comparative Study of Existing Methods for Localization in WSN

S.No	Existing Method	Performance Metrics	Limitations
1	Global Position System (GPS)	This method tested using different parameters like location error, Complexity and energy consumption	This method gives more error, more complexity, and more energy consumption.
2	Multilateration	MSE is calculated in this method	This method gives more error.
3	Nearest three references	MSE is calculated in this method	MSE depends on the accuracy of nearest reference nodes.
4	Distributed least mean square (DLMS)	MSE and complexity are calculated by varying number of nodes in this method.	This method gives high MSE.
5	Distributed recursive least-squares (DRLS)	MSE and complexity are calculated by varying number of nodes.	This method gives high complexity.
6	Particle swarm optimization (PSO)	MSE and complexity is calculated using centralized and distributed approach.	This method gives more MSE in indoor environment due to multipath effects.
7	Kalman filter (KF)	MSE is calculated for tracking of mobile nodes.	This method does not work well in a nonlinear environment.
8	Extended Kalman filter (EKF)	MSE is calculated for tracking of mobile nodes in a nonlinear environment.	This method gives more MSE in a high-noise environment.

algorithm has a low complexity, implantation is simple, has a high degree of convergence [19], and is visible for WSN localization. A distributed PSO algorithm based on the probabilistic distribution of ranging error is proposed to increase the performance and accuracy of PSO-based localization approaches [20–23]. The suggested goal function assesses particle fitness. It tries to locate further unknown nodes in a high-accuracy search space [7]. The updated particles position can be mathematically modeled according to Equations (11.1) and (11.2).

$$V_i^{k+1} = wV_i^k + c_1 r_1 \left(p_{best} - x_i^{\ k} \right) + c_2 r_2 \left(g_{best} - x_i^{\ k} \right) \tag{11.1}$$

$$X_i^{k+1} = X_i^{\ k} + V_i^{k+1} \tag{11.2}$$

In the following calculations, the modified component location can be arithmetically modeled via Equations (11.1) and (11.2), where V_{ik}, X_{ik} are speed and current node I location are at the iterations k, C_1 and C_2 are a constant random number (r_1 and r_2), distributed uniformly in (0,1), and w, the inertia, is the weight needed to monitor the search distance. W is usually set to decrease in line with the progression p_{best}, g_{best} are and best of the particles in the world.

Let (x, y) be unknown node U coordinates and (x_i, y_i) be location of its neighboring A_i of U anchor node i^{th} (i = 1, 2,..., m) and dividing by Equation (11.3), the distance measured between U and A_i, gives:

$$F(x,y) = \frac{1}{M} \sum_{i=1}^{M} \sqrt{(x - x_i)^2 + (y - y_i)^2} - \hat{d}_i \tag{11.3}$$

The location function f is known as all range-based localization methods using PSO do not take into consideration the stochastic range error distribution function. If the real interval between U and Ai is an Equation 11.4 natural distribution.

$$d_i \sim N\left[\hat{d}_i, \left(\hat{d}_i \delta \right)^2 \right] \tag{11.4}$$

where $\delta > 0$

$$\sum_{i=1}^{M} d_i \sim \prod_{i=1}^{M} N\left[\hat{d}_i, \left(\hat{d}_i \delta \right)^2 \right] \tag{11.5}$$

The ranging error distributions between various nodes are independent, as seen in Equation (11.5).

In order to boost position precision, Equation (11.6) with the error distribution is used.

$$f(x,y) = \frac{1}{M} \sum_{i=1}^{M} \frac{\sqrt{(x - x_i)^2 + (y - y_i)^2} - \hat{d}_i}{\left(\hat{d}_i \delta \right)^2} \tag{11.6}$$

TABLE 11.2
Simulation Setup

S.No	Parameter	Value
1	Simulation area	500×500 m
2	No. of anchor nodes	13
3	No. of unknown nodes	100
4	Transmission range	20 m
5	Initial energy	5 J
6	Transmission power	2
7	Path loss exponent	2
8	Threshold value (conservative approach)	0.08
9	Threshold value (aggressive approach)	0.05

11.3.1 SIMULATION AND RESULTS ANALYSIS

11.3.1.1 Simulation Setup

MATLAB® simulates and checks the suggested solution CDPSO with censoring. There are 100 randomly located agent nodes and 13 anchor nodes with known positions within the 500×500 m region. The simulation parameters and values are shown in Table 11.2.

11.3.1.2 Results Analysis

The plotted graph for anchor vs. processing time is seen in Figure 11.2. The cycle time also increases with the node of anchor nodes. Therefore, CRB picks the best nodes of reference. The graph showing the number of anchor nodes vs. position errors is seen in Figure 11.3. With the increase of the anchor nodes, location errors are reduced. A small and constant error is observed in the number of anchor nodes greater than or equal to 4. The performance of the proposed method cooperative distributed PSO (CDPSO) algorithm was tested with different parameters, and results were analyzed. In a cautious method, Figure 11.4 depicts a graph between position error and CDF (RTx = 0.08). The distributed PSO algorithm with CRB in terms of combined distributed PSO is in conjunction with such algorithms as GPS, LMS, RLS, and PSO (CDF). Compared to other algorithms, the CDPSO CRB produces better performance. The proposed approach discards the transmission of incorrect node location information while selecting the nodes that deliver precision results. The erroneous position knowledge of agent nodes is often used as anchor nodes (Figure 11.5).

The CDPSO CDF is therefore degraded. For RLS, LMS is decreased as a result of the smaller number of pre-fixed nodes involved in the location operation. The non-line of sight and multipath results of GPS are bad. GPS is poor. Figure 11.4 displays the graph of various hostile methods (RTx = 0.05) between the location error and CDF. The CDPSO CRB algorithm was compared to PSO, RLS, LMS, and GPS algorithms. The comparison showed that the CDC CDPSO with the CRB works well because it chooses optimal references with high precision, discards position data from incorrect nodes, and locates the nodes.

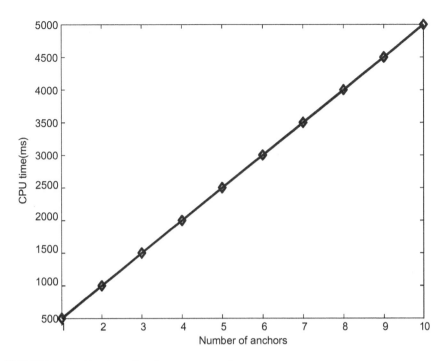

FIGURE 11.2   Number of anchors vs. processing.

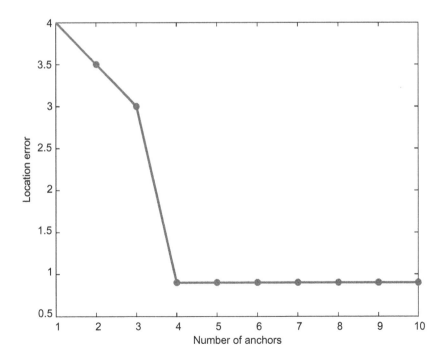

FIGURE 11.3   Number of anchor nodes vs. time location error.

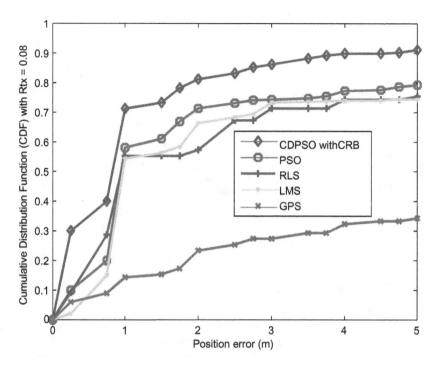

FIGURE 11.4 Conservative approach.

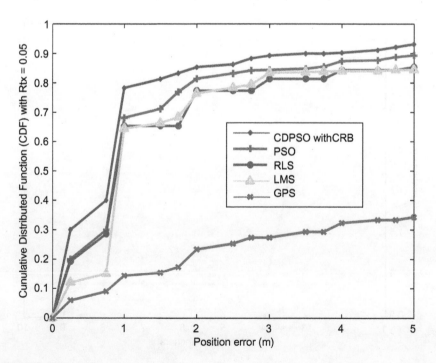

FIGURE 11.5 Aggressive approach.

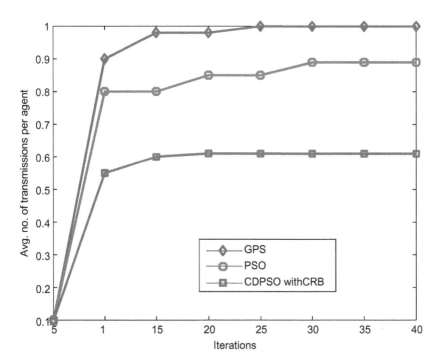

FIGURE 11.6 Average number of transmissions.

In the case of PSO, the localization method includes nodes with incorrect position knowledge. Thus, similar to other algorithms, the CDF is degraded. RLS and LMS performance is low due only to the position of previously installed nodes. GPS efficiency, due to NLOS, noise and multipath effects, is less compared to other algorithms.

Figure 11.6 displays a graph showing the total number of transmissions every five iterations. A larger number of incorrect nodes are excluded from the CDPSO complexity with CRB. The locomotive mechanism thus involves the least number of nodes. Complexity is also decreased. PSO is more complicated due to any node that is participated in the localization phase in the transmission spectrum as a reference node. This also raises the number of transmissions to identify individual agent nodes. In GPS signals, noise, multipurpose, and computer sophistication are more influenced.

The graph between the number of locations for every five MSE iterations is shown in Figure 11.7. CDPSO's CRB algorithm was compared to other PSO, RLS, LMS and GPS algorithms. The CDPSO algorithm works better in this comparison because nodes with correct data engage only in the localization process. In the event of PSO faulty nodes, the output will also be degraded. The efficiency of the LMS and RLS algorithms is low because of the smallest number of LMS and SNR references. In the event of GPS, noise and multipath effects are impaired, so MSE is much more important compared to other algorithms.

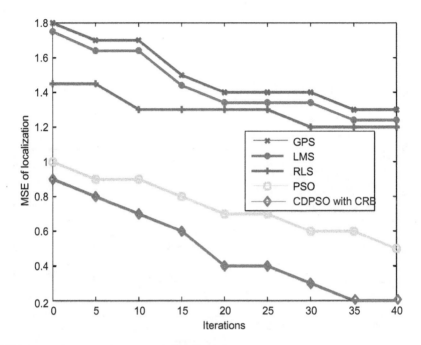

FIGURE 11.7   Mean square error of localization.

11.3.2   PSO ASSISTED AKF ALGORITHM

Kalman filter is a repetitive filter that is commonly used in linear environments for position estimation. Kalman filter extended (EKF) is an amendment in a nonlinear Kalman filter. The multipath effects EKF does not provide the optimal solution indoor atmosphere due to noise. Particle swarm optimization (PSO) is a population-based search algorithm formulated on swarm intelligence, e.g., birds' social conduct, bees, or a fish academy. The suggestion for optimal reference came in this PSO-aided Kalman filter (PSO-AKF). Figure 11.8 shows the PSO-AKF algorithm. Grade of range of divergence (ROD) is the difference between $\hat{C}_{vk}$ and C_{vk} gives the trace of innovation covariance matrix. This element is used to define the variations or adaptations in Equation (11.7) for adaptive filtering.

$$FIT = ROD = \frac{tr\left(\hat{C}_{vk}\right)}{tr\left(C_{vk}\right)} \tag{11.7}$$

ROD is used to determine fitness or to aid the EKF algorithm in determining the valuation threshold in a nonlinear scenario. ROD is an innovation matrix that is used to determine value uncertainty. If ROD is more than the stated threshold value, the Q_k and R_k by scaling factor will be changed. The ROD fitness function parameter is used to calculate a PSO-aided AKF (FIT). Using PSO to iteratively tune the FIT parameter assures optimization.

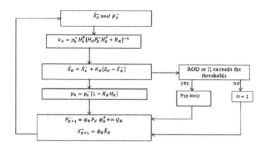

FIGURE 11.8 PSO-assisted AKF.

11.3.3 SIMULATION AND RESULTS

Four cooperative WSN simulation studies with various tests were performed. Proposed PSO-AKF algorithm compared to various algorithms such as LS, KF, EKF. Place errors Figure 11.9 CDF of different algorithms. Accuracy of the localization method for location measurement error for various values was evaluated using Cumulative Distribution (CDF) algorithm.

AKF supported by PSO Better efficiency relative to EKF (PSO without EKF assisted) Comparison of multiple ROD algorithms as seen in Figure 11.10. Method suggested The distribution of PSO-AKF to CRB increases cumulative distribution error. The location is compared to other algorithms

The PSO-supported AKF distributed is the least, as seen in Figure 11.11, and the cooperative distributed PSO-AKF with CRLB does best in range of divergence (ROD) localization as seen in Figure 11.12 relative to other systems.

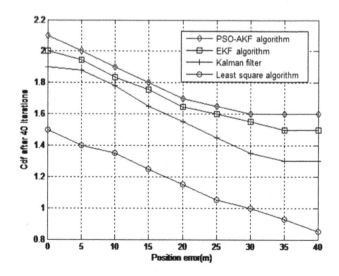

FIGURE 11.9 Performance comparison of different algorithms.

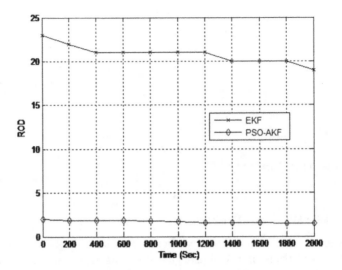

FIGURE 11.10   PSO assistant and non-PSO ROD comparative analysis.

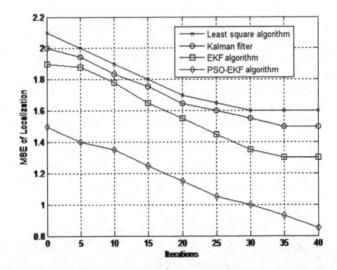

FIGURE 11.11   A comparison of various methods for mean square error of localization in relation to the number of iterations.

This section suggests the position of the cooperative WSN with maximal references, using dispersed PSO-aided AKF. The cooperative distributed assisted PSO-AKF algorithm CRB is applied as comparison selection tool. In applications of position estimation, Kalman filter is a recursive linear filter. Distributed PSO-AKF distributed with CRB improves on cumulative error of the distribution function. In contrast to other schemes, complexity was minimized by means of the proposed form. A cooperative dispersed PSO-AKF with the CRLB works in a location which is less significant than other schemes with a square error and a ROD.

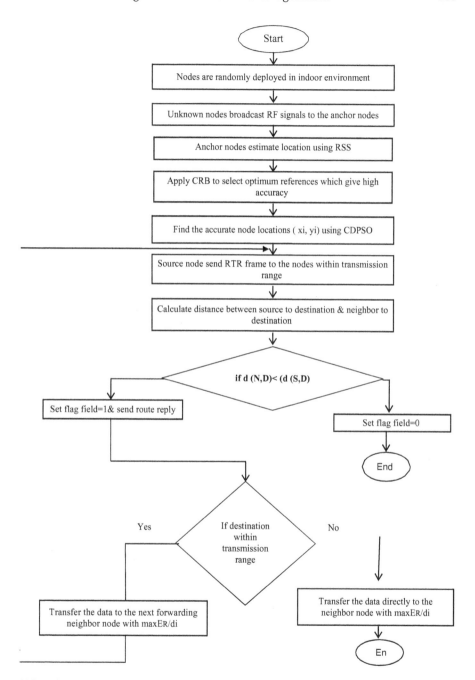

FIGURE 11.12 Flow chart for CDPSO routing algorithm.

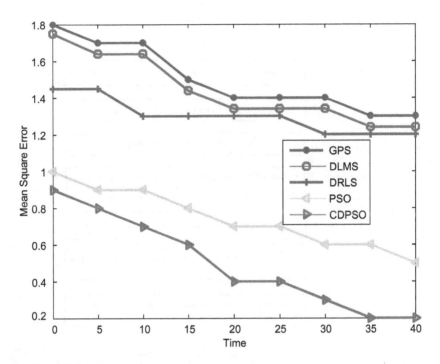

FIGURE 11.13   Mean square error vs time.

11.3.4   CDPSO LOCALIZED ROUTING WITH OPTIMUM REFERENCES

This section proposes a localized routing algorithm with optimal references for the cooperative distributed particle swarm optimization. Reference nodes are chosen in the proposed solution with correct position information. The algorithm Cramer Rao Bound selects the reference nodes that provide exact position knowledge. CPDSO is used to locate the accurate nodes. The proposed mechanism with different criteria was applied and evaluated. The exact location of the suggested algorithm nodes using the optimal references distributed PSO algorithm indicates better results as regards mean square error (MSE). This exact location is used for transferring data from source to destination. The findings of a simulation show that dispersed mutual, located PSO routing performs better with optimal relation in contrast to the PSO algorithm, the DLMS (distributed least mean squares), and the DRLS algorithm. Figure 11.12 shows a flow map for the energy-efficient routing suggested using a distributed PSO location.

11.3.5   SIMULATION RESULTS AND ANALYSIS

MATLAB® has been used for simulation tests. These parameters are checked for CDPSO, PSO, DRLS, DLMS, and GPS algorithms. Compared to other algorithms, CDPSO's suggested routing algorithm performs better. The effects of several parameters such as PDR, remote capacity, throughput, and mean square error are seen in Figures 11.13–11.16. Table 11.3 summarizes the performance of the various algorithms.

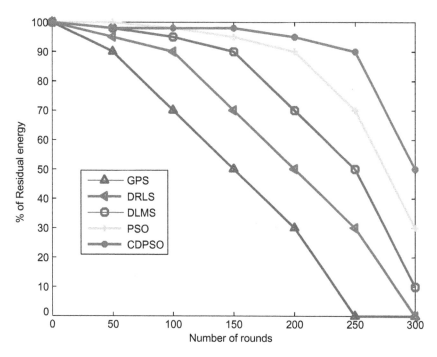

FIGURE 11.14 Residual energy vs No. of rounds.

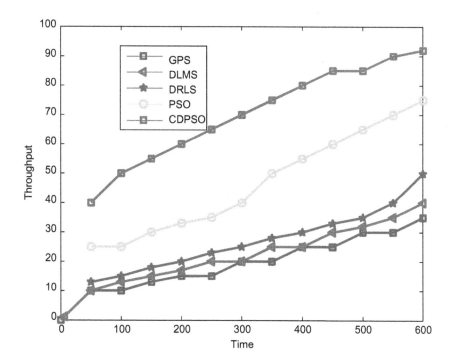

FIGURE 11.15 Comparison of throughput.

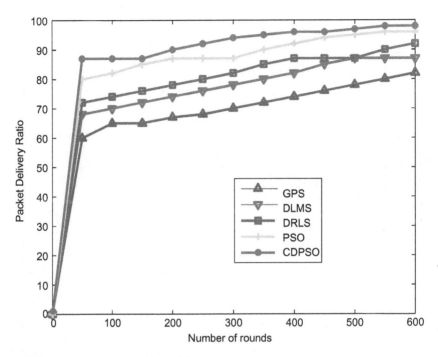

FIGURE 11.16   Packet delivery ratio vs. time.

TABLE 11.3
Performance Comparison of Different Algorithms

Parameter	CDPSO	PSO	DRLS	DLMS	GPS
Nodes that are still working	78	65	58	48	36
MSE	0.7	1.3	1.4	1.5	1.7
Avg. delay	20	26	34	41	50
Residual energy	89	83	73	62	48

In this section energy-efficient routing in WSN using CDPSO Localization with optimal references is proposed. Precise location of nodes can be determined using CRB.

11.3.6   LOCATION TRACKING OF PATIENTS USING PSO-AKF

In hospitals, patient surveillance and recording is crucial. In hospital environments, it is difficult to locate and monitor Alzheimer's patients. Persons who suffer from Alzheimer's disease were trained to find and trace their movements in hospital environments and relate such movements to a WSN. WSN nodes are used when an Alzheimer's sufferer is wired to a network with a sensor node. The first step involves RSS. Due to noise effect signals strength, the suggested PSO-AKF for location information and monitoring of patient movements with out-of-sight error. The first steps are to improve location information.

For position monitoring and tracking of patient movement, an improved localization algorithm with particle swarm optimization is proposed for this segment. The extended Kalman filter (PSO-AKF) measures the positioning node PSO-assisted AKF in the planned enhanced local routing. The first part fixes the PSO-AKF spot. The theme is the establishment of the path and destinations. PSO is a tool for optimizing the bird group's behavior. The PSO is used to tune assist covariance of innovation sequence in this proposed algorithm.

11.3.7 SIMULATION RESULTS AND ANALYSIS

The approach proposed is validated and used in the hospital community for the application of monitoring patients. The MATLAB® simulation results show that Kalman filters, helped by particle swarm optimization, are doing better as compared to other algorithms with regard to position errors, precise monitoring, delay minimization, and minimization of complexity (Figures 11.17–11.20 and Table 11.4).

A way of efficiently pursuing the WSN indoor application is suggested in this process. The suggested approach uses the Kalman extended adaptive filter (PSO-AKF) particle swarm optimization to locate node locations. This location information is used in indoor applications to map and route data. In the application of monitoring patients, the proposed approach is checked and used. The MATLAB® simulation results show that Kalman filters, helped with PSO, do better as compared to other algorithms with regard to position errors, precise monitoring, delay minimization, and minimization of complexity.

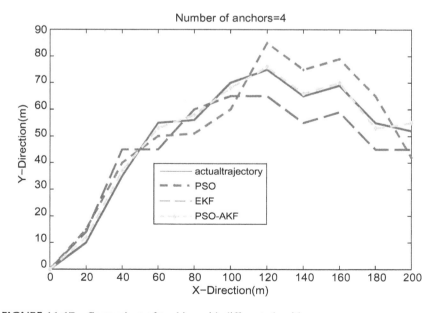

FIGURE 11.17 Comparison of tracking with different algorithms.

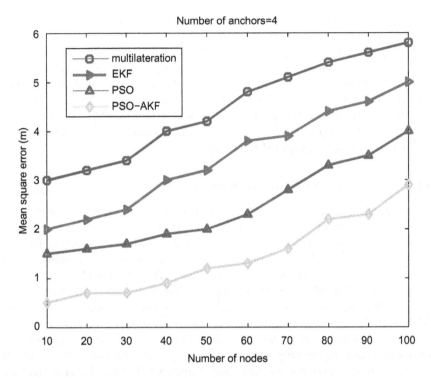

FIGURE 11.18 Comparison of mean square error.

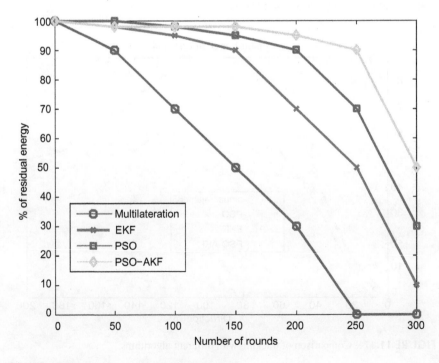

FIGURE 11.19 Comparison of residual energy.

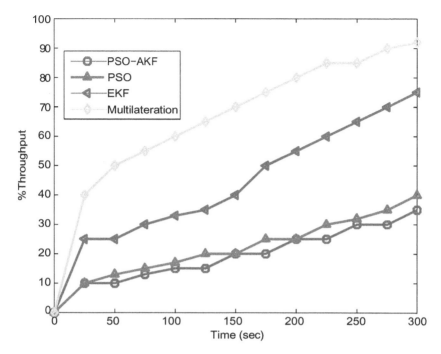

FIGURE 11.20   Comparison of throughput.

TABLE 11.4
Performance Comparison of Different Algorithms

Parameter	Multilateration	PSO	EKF	PSO-AKF
Alive nodes	20	31.6	48.3	61.6
Throughput (Mbps)	22.5	24.3	47.5	72.8
Residual energy	20.3	32.3	52.5	62.6
PDR	73	79.5	90.5	93.6
Delay (Sec)	5.5	4.5	2.9	2.2

11.4   CONCLUSION

The precise positioning of the node was suggested for the cooperative dispersed PSO algorithm CRB (CDPSO-CRB). With the support of anchor nodes, this algorithm finds the exact node location. To approximate minimum difference, CRB shall be used to exclude incorrect ties. Place precision and lower complexity are the advantages of the proposed approach compared with other algorithms. The optimal reference suggested for optimizing node location accuracy is the particle swarm optimization adaptive comprehensive Kalman filter (PSO-AKF). The CRB algorithm is extended to a cooperative distributed AKF-assisted algorithm as a comparison selection process. PSO is used to tune or support covariance of the creativity series in this proposed algorithm. The method suggested that the cumulative distribution function of distributed PSO-AKF with CRB is stronger. In contrast to other schemes, complexity was minimized by means of the proposed form. Cooperative distributed PSO-AKF with CRB increases localization efficiency in contrast to other schemes with lowest mean square error and ROD.

It is suggested to find CDPSO with CRB. CRB chooses the high-precision nodes. This algorithm selects reference nodes which offer the position information minimal variance and high accuracy. CDPSO seeks precise node sites. The data is transmitted from source to destination using this position data. The computer complexity suggested mechanism offers the least MSE. Nodes used in the routing process are located. In terms of network life, output, PDR, residual power, and average delay, the proposed mechanism performs better compared with the other algorithms such as GPS, DLMS, DRLS, and PSO.

REFERENCES

1. Kulkarni, R. V., Forster, A., Venayagamoorthy, G. K. (2011). *IEEE Communications Surveys Tutorials*, vol. 13, no. 1, pp. 68–96, First Quarter 2011. doi:10.1109/SURV.2011.040310.00002.
2. Rezazadeh, J., et al., (2014). "*IEEE Sensors Journal*, vol. 14, no. 9, pp. 3052–3064.
3. Seco-Granados, G., Lopez-Salcedo, J., Jimenez-Banos, D. (2012). *Signal Processing Magazine, IEEE*, vol. 29, no. 2, pp. 108–131.
4. Klonovs, J., Haque, M. A., Krueger, V., Nasrollahi, K., Andersen-Ranberg, K., Moeslund, T. B., Spaich, E. G. (2016). In J. Klonovs et al. (eds.), *Distributed Computing and Monitoring Technologies for Older Patients* (pp. 49–84). Springer International Publishing, Chicago.
5. Seco-Granados, G., Lopez-Salcedo, J., Jimenez-Banos, D. (2012). *Signal Processing Magazine, IEEE*, vol. 29, no. 2, pp. 108–131.
6. Sahoo, B. M., Amgoth, T., Pandey, H. M. (2020). *Ad Hoc Networks*, vol. 106, 102237.
7. Sha, K., Gehlot, J., Greve, R. (2013). *Wireless Personal Communications*, vol. 70, no. 2, 807–829.
8. Sahoo, B. M., Pandey, H. M., Amgoth, T. (2020). *Swarm and Evolutionary Computation*, vol. 60, 100772.
9. Pagano, S., Peirani, S., Valle, M. (2015). *Wireless Sensor Systems, IET*, vol. 5, no. 5, 243–249.
10. Wang, W., Ma, H., Wang, Y., Fu, M. (2015). *Ad Hoc Networks*, vol. 25, pp. 1–15.
11. Ma'sum, M. A., Rahmah, N., Sanabila, H. R., Wisesa, H. A., Jatmiko, W. (2015). *Automatic fetal head approximation using Particle Swarm Optimization based Gaussian Elliptical Path*. In *2015 International Symposium on Micro-NanoMechatronics and Human Science (MHS)*, November 23 (pp. 1–6). IEEE.
12. Miraswan, K. J., Maulidevi, N. U. (2015). *Particle swarm optimization and fuzzy logic control in gas leakage detector mobile robot*. In *2015 International Conference on Automation, Cognitive Science, Optics, Micro Electro-Mechanical System, and Information Technology (ICACOMIT)*, October 29 (pp. 150–155). IEEE.
13. Esmin, A. A., Coelho, R. A., Matwin, S. (2015). *Artificial Intelligence Review*, vol. 44, no. 1, pp. 23–45.

14. Liu, J., Chen, Y., Chen, X., Ding, J., Chowdhury, K. R., Hu, Q., Wang, S. (2013). *Journal for Control, Measurement, Electronics, Computing and Communications*, vol. 54, no. 4, 438–447.
15. Yao, Y., Jiang, N. (2015). *Computer Networks*, vol. 86, pp. 57–75.
16. Popescu, A. M., Salman, N., Kemp, A. H. (2013). *Wireless Communications Letters, IEEE*, vol. 2, no. 2, pp. 203–206.
17. Bours, A., Cetin, E., Dempster, A. G. (2014). *Electronics Letters*, vol. 50, no. 19, pp. 1391–1393, September 11. doi:10.1049/el.2014.2248.
18. Lakshmanan, L., Tomar, D. C. (2014). *American Journal of Applied Sciences*, vol. 11, no. (3), pp. 520.
19. Zungeru, A. M., Ang, L.-M., Seng, K. (2012). *Journal of Network and Computer Applications*, vol. 35, no. 5, pp. 1508–1536.
20. Zheng, J., Qin, T., Wu, J., Wan, L. (2016). *Eighth International Conference on Advanced Computational Intelligence (ICACI), Chiang Mai*, vol. 2016, pp. 41–44. doi:10.1109/ICACI.2016.7449800.
21. Abdolee, R., Champagne, B. (2011). *International Conference on Distributed Computing in Sensor Systems and Workshops (DCOSS), Barcelona*, 2011, pp. 1–6. doi:10.1109/DCOSS.2011.5982212.
22. Mateos, G., Schizas, I. D., Giannakis, G. B. (2009). *Signal Processing, IEEE Transactions On*, vol. 57, no. 11, pp. 4583–4588.
23. Bertrand, A., Moonen, M., Sayed, A. H. (2011). *Signal Processing, IEEE Transactions On*, vol. 59, no. 11, 5212–5224.

12 A Review on Defense Strategy Security Mechanism for Sensor Network

Amara SA L G Gopala Gupta

Koneru Lakshmaiah Education Foundation Guntur, Andhra Pradesh, India

G. Syam Prasad

Narasaraopeta Engineering College, Narasaraopeta, Guntur, Andhra Pradesh, India

CONTENTS

12.1 INTRODUCTION

At present, the wireless sensor network (WSN) has been widely adopted in a vast range of applications. WSN provides a drastic range of connection diversified in nature, this put a high risk on security [1]. To withstand security risks, attack graphs of sensor network indicate the behavior and related intruders' vulnerabilities and

DOI: 10.1201/9781003145028-12

show all possible attack paths in sensor network intruders. Each path is considered as an attack scenario which states the relationship between exploits and attack sequence. By using this graph, security administrators discover the network risks and security threats. Based on requirements to avoid dangerous events, proper security defense measures are selected [2,3]. To improve security mechanisms in sensor network defense, a strategy-based map is constructed in an attack graph with inclusion of attack library. The distinction between a defensive strategy map and an attacker graph is that the former provides outcomes from the defender's perspective, which is more comprehensive and significant than the attacker's perspective. The ideal defense [4] method prevents all attack behaviors but considers organizational resource constraints which use limited resources to take reasonable decisions. For security administrators, defense cost is the major factor [5].

The defense strategy in a sensor network is improved through inclusion of the optimization technique in attack graphs. That optimization technique adopts a binary approach with inclusion of local extremum and minimal convergence, but it is subjected to premature convergence. The premature convergence problem in the binary approach is eliminated through path extraction and insertion. This implies that a game theory-based security mechanism is considered an effective security scheme for WSN. Hence, this chapter reviews the security mechanism based on game theory.

The chapter is organized as follows: Section 12.1 presents a general description about challenges in WSN security. Section 12.2 discusses WSNs' game theory, while Section 12.3 gives the classification of the game theory method. Section 12.4 summarizes the class of game, game solution methods, and energy-saving techniques utilized, and Section 12.5 describes survey conclusions. On the whole, this chapter presents the impact and contribution of game theory on WSN security.

12.2 GAME THEORY IN WIRELESS SENSOR NETWORKS

Game theory is widely applied in an intelligent system, especially in a challenging environment. The application of game theory to the modern wireless communication network is stated in the book *Game Theory for Wireless Engineers* [6]. Usually, game theory is observed as an effective approach when integrated with other types of wireless communication network. The integration of game theory with another network is performed due to consideration of following constraints:

1. The WSN network needs to be distributed fully or partially.
2. WSN nodes are fixed and homogeneous and have limited battery life and hardware resource constraints.

To withstand those factors and increase the lifetime of the network with increased quality of service (QoS), game theory is adopted. However, recent research has concentrated on selection of game model for a specific problem. Further, the developed game model needs to provide guaranteed convergence to achieve the desirable solution in an appropriate amount of time [7]. Figure 12.1 presents the classification of the game theory model, and its description is presented in the following sections.

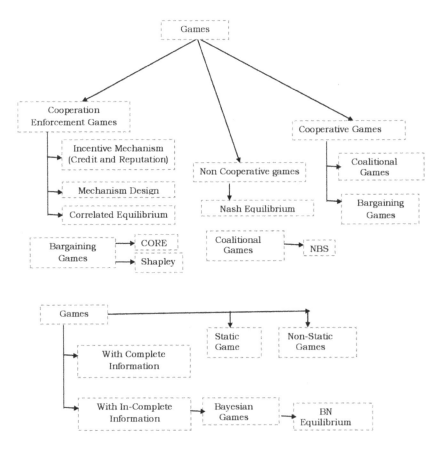

FIGURE 12.1   Classification of game theory.

12.2.1 Classification of Games

In the literature, games are classified into two classes: cooperative and noncooperative games [8]. Some research utilizes another type of game theory, defined as cooperation enforcement games. It is observed that there are numerous articles under this class, thus helping the reader to understand game theory models in a better way. Every class with subsections describes various concepts with examples, as presented in Table 12.1.

12.2.1.1 Noncooperative Games

Noncooperative game theory focuses on individual player utility instead of the complete network. In this game, every node in the network acts as selfish due to the following reasons:

1. Data forwarding impacts on individual node resources.
2. Data forwarding affects the connectivity of the network.

TABLE 12.1
Game Theory in WSN

Article	Class of Game	Method Used
Byun et al. [9]	Cooperative	Increasing number of bits transmitted per watt of consumed power.
Chai et al. [10]	Noncooperative	In energy consumption model, computational and sensing energy is considered.
Hakim and Jayaweera [11]	Cooperative	In fusion center WSNs, fair allocation power in collaboration nodes.
Hao et al. [12]	Noncooperative	To minimize energy consumption and interference, joint channel allocation, and energy consumption.
Luo et al. [13]	Noncooperative	For quick convergence to NE point, game with energy-efficient convergence method.
Sengupta et al. [6]	Noncooperative	Based on various channel condition, distributed power control method.
Tsuo et al. [7]	Noncooperative	With reduced overhead, energy-efficient power control.
Tushar et al. [8]	Noncooperative	To achieve target SINR; optimization of transmission power.
Zhang et al. [14]	Noncooperative	For residual energy of nodes, consider SINR method.

The connectivity scenario considered in noncooperative game is based on the Forwarder's Dilemma [10]. To achieve an effective solution for a noncooperative game, Nash equilibrium (NE) is considered as the central solution, without deviating single player performance. The strategy vector is represented as with NE of player i, and its corresponding strategy is denoted as s_i. The game theory model for non-cooperative model is defined as follows:

$$u_i\left(s_i, s_{-1}\right) \geq u_i\left(s_i', s_{-1}\right)$$

NE denotes a definite stable point of operation which is powerful to unilateral deviations. This theorem explains that each finite game in strategic form has either a mixed or pure NE [9]. In the Forwarder's Dilemma game, the NE point is one in which every node drops from other nodes' packets. For a game, NE is in pure strategy, while every player plays his own strategy. Each player chooses certain probability distribution in strategy sets, which is known as mixed strategy. Using Pareto optimality, one method for finding the NE point in game and by comparing the strategy profiles, a solution maximizing the utilities of both players can be confirmed [11]. In the Forwarder's Dilemma, the strategy profile is Pareto-optimal but not NE when both nodes forward packets. Potential game having potential function is used to show the best response dynamic coverage to point of equilibrium. Potential games have many remarkable properties.

12.2.1.2 Cooperative Games

The cooperative game mechanism is achieved through a high range of cooperation between players. The cooperative games are classified in two categories as follows:

1. Reputation-based approach.
2. Credit-based approach

In a reputation-based mechanism, every node gains some value from other nodes by reputation process. In the case of a credit-based mechanism, every node receives a certain amount of payment from relay nodes for forwarded packets. Bargaining games, a broadly applicable concept for cooperative games, examines the case where two or more players have to choose any one of multiple possible outcomes of the joint collaboration. For example, consider players agreeing to a fair sharing of resources within a cluster. When this agreement is reached, both are beneficial rather than playing without cooperation. The solution thus obtained is known as the Nash Bargaining Solution (NBS) [15], where action was not taken by one individual without the consent of the other. With cooperative games, the difficulty is that its nodes need to accomplish some agreements and additional computations among one another.

12.2.1.3   Cooperation Enforcement Games

Enforcement model performance is based on the cooperation of nodes in which a strategic solution is achieved by means of desirable outcome. This game model is implemented in a multiple-resource admin platform. This model utilizes the Vickery–Clarke–Groves model for processing. Based on the information obtained from the owner, the Bayesian model is applied for minimal game performance. To evaluate the coordinator of the game, correlated equilibrium is preferred to perform external correlation [16]. In the enforcement model, two players are considered as driving point at same time. In this game, players are considered as strategically independent factors. Correlated equilibrium adopts computational strategy for deriving a polynomial solution. This identifies optimal correlation value based on observation and controlling process in the game model [17].

12.2.1.4   Other Classification

Game theory model functionality is based on the consideration of certain criteria like whether the game is static or dynamic. In the static game model, it is assumed that only one time step is played for simultaneous process. Some researchers utilize game theory in WSN for energy saving in which noncooperative NE operation is based on repeated game process which provides incomplete information. The proposed model is based on the certain control mechanism with consideration of a different channel [6]. Also, noncooperative NE [18] is utilized on WSN for achieving residual energy. Another research, noncooperative Bayesian Nash equilibrium (BNE), is adopted for energy-efficient performance with minimal overhead [19]. Effective optimization transmission power is achieved through consideration of target SINR [20]. To resolve convergence optimization, the noncooperative algorithm is computed for reducing energy consumed and sensing [21]. In [22], the noncooperative ordinal potential game is applied for channel allocation for power control for reducing energy consumption. With Cooperative Coalition for maximizing the number of bits processed with increased lifetime, the Deferred Acceptance Procedure (DAP) is developed by [13]. In [12], for optimal power transmission Shapley is adopted for cooperative coalition for total source cost and tradeoff for achieving energy efficiency and reduced end-to-end delay. In Table 12.2 different types of games and their concepts are presented.

TABLE 12.2
Game Theory for Security in WSN

Concepts of Game Theory	Used By
NE-best response dynamics	Chai et al. [10], Hao et al. [12], Jiang et al. [15], Luo et al. [13], Sengupta et al. [6], Tushar et al. [8], and Zhu and Martinez [17]
Repeated game	Abid and Boudriga [20], Bharathi and Kumar [21], Pandana et al. [23], Sengupta et al. [6], Wei et al. [22], Zhang et al. [14], Zhao et al. [16], and Zhu and Martinez [17]
Incomplete information	Abid and Boudriga [20], Dai et al. [24], Hao et al. [12], Ren and Meng [18], Sengupta et al. [6], and Tsuo et al. [7]
Bayesian NE	Pandana et al. [23], Ren and Meng [18], and Tsuo et al. [7]
Dynamic game	Ren and Meng [18] and Zhu et al. [14]
Pure strategy NE	Bharathi and Kumar [21] and Zhang et al. [14]
Pareto optimality	Ginde et al. [25], and Hao et al. [12], and Niyato et al. [26]
Mechanism design	Dai et al. [24] and Zhang et al. [14]
Evolutionary game	Jiang et al. [15] and Zhang et al. [14]

12.3 SECURITY DEFENSE STRATEGY ATTACK GRAPH

The identification of a critical attack scenario in sensor nodes is estimated based on the consideration path as attack in the set. At present, the attack graph utilizes a binary optimization approach for reducing maximal convergence. The application of defense strategy in different attack scenario are stated in what follows.

12.3.1 GAME THEORY FOR SYBIL ATTACK

Douceur discussed about Sybil attack and defense using a method based on trusted certification of identity. In this method, the authority recognizes the entity and provides unique digital signatures. Cryptography is used for entity authentication. For all kinds of network, this method was generalized. Due to implementation overhead, it is costly to execute in large networks. Various methods are presented in [24] to defend against Sybil attack and introduced a method based on radio resource entity testing. Energy levels as well as storage capacity were evaluated and compared. Identity with more radio resources is known as Sybil identity. Quantitative evaluation of Random Key Pre Distribution (RKPD), entity registration, and position verification are other defense methods that were discussed. Due to use or testing multiple radio links, battery power of the appropriate node weakens, and malicious nodes are not considered. In WSNs [14] presents Sequential Hypothesis Testing (SHT) to defend against Sybil attack. In WSNs, [27] developed a Sybil defense method called Received Signal Strength of Identities (RSSI) which stands for power existing at the input of the receiver node. In RSSI, a message sent by various Sybil nodes identifies the malicious node having identical signal strength at the input receiver node. Sybil identity is estimated using these values. In WSNs, due to fading signal, mobility problems and false positives, this method seems to be less reliable. These papers are not fully based on game theoretical method, but it gives a basic understanding of mitigation and attack detection methods employed. With the help of game theory,

[26] introduced zone trust approach using sequential hypothesis testing (SHT). The whole network is partitioned as zones in which SHT is utilized to determine the suspected Sybil region. With game theoretical analysis, compromised nodes were detected. For defender as well as attacker, NE strategies are defined.

For mobile ad hoc networks (MANETs), [28] designed a game theory-based reputation method. Between attacker and defender, this method makes attack costly in every stage. Based on the game theoretical problem, [29] presents a protocol. A reputation system was introduced in which nodes were the vertices joined by edges. In a game, mixed strategy NE is described which suggests that only a good node is said to be a Sybil node with mixed strategy profile for exposing Sybil nodes and preventing malignancy coalition in the network. In [30], an incentive-based Sybil attack defense termed as informant was designed for ad hoc networks. It contains three players, namely Detective, Target, and Informant. Reward is given to the informant in form of payment if Sybil identities controlled by it are disclosed. To evaluate minimum award value, the Dutch auction technique is utilized. For game, NE is not evaluated. The protocol coined tempts the attacker to commit false claims as a Sybil identity with the motive of gaining more rewards, which is harmful to Sybil defense.

12.3.2 Defense Strategy for Denial of Service (DDoS)

DDoS attacks occur in various internet network layers. Powerful defense methods like detection, prevention and response methods were used in various layers. In [2], general categorization approach to classify the layer-based defense mechanisms were presented which included transport/network and application layers.

12.3.3 Defense Mechanisms of Transport/Network Layer

A defense mechanism, based on the transport or network layer performance is stated in to consideration of different groups such as UDP, TCP, and ICMP protocols.

12.3.3.1 Source-Based Mechanism

Source-based defense methods along with its properties are described and compared in this section. These methods were created and utilized for defense in a source region. This method was used for detecting anomalous flows and passing packets as well as performing defense actions like rate limiting and filtering [3]. In [4], the Tabulated Online Packet Statistics (TOPS) monitoring approach was introduced to detect as well as filter bandwidth DoS attacks, which utilized heuristic rules to estimate traffic. For monitoring space and IP address domain, fixed compact tables were utilized to detect packet flow imbalance. Despite a traditional firewall, in [5], a reverse firewall was developed by MANET to filter outgoing packets. For transmitter engine, this method restricted the speed of packet transmission. Inside the network, it is significant in mitigating DDoS attack effects. In [6], a method based on quantitative measurements was designed to detect DDoS attacks where two proportion factors are obtained to compromise the host, which significantly impacts on the deviating traffic feature. For detecting the subtle DDoS anomaly in monitors nearer to the attack source, a multi-stage detection method incorporating Network Traffic State (NTS), malicious address

TABLE 12.3

Defense Strategy and Techniques

Author	Defense Game Theory	Technique
Bonnet [29]	Coalition and NE	Optimal strategies
Fallah [28]	Cooperative	Game theory based reputation mechanism
Koutrouli and Tsalgatidou [27]	Cooperative	Defense Mechanisms
Kumar et al. [1]	Cooperative	Sequential Hypothesis Testing (SHT)
Vasudeva and Sood [3]	Noncooperative	Received Signal Strength of Identity (RSSI)

extraction engine, and fine-grained singularity detection was designed. NTS prediction was used to detect the network deviation rate at the monitoring point. In Table 12.3, techniques and type of game theory applied are presented.

12.3.3.2 Routing Defense Mechanism Game Theory for Sensor

The deployment of game theory on sensor nodes is subject to certain routing mechanisms. Based on the defense mechanism, attack nodes are detected and isolated. The attack detection mechanism based on game theory is categorized as follows:

1. Probabilistic based mechanism.
2. Ferry-based detection.
3. Reputation-based detection.
4. reference-based detection.

In case of the probabilistic detection approach, the trusted authority collects the information about nodes [31]. In [32], ferry-based attack detection is performed with consideration of mobile nodes in the network. In Figure 12.2, a defense mechanism adopted for sensor node is presented.

To achieve attack detection accuracy, mutual correlation is achieved through ferry-based detection. However, the proposed scheme reduces network efficiency with increased cost [33]. In [34], a reputation-based dynamic approach is applied for estimation of reputation in node. The information related to all nodes is gathered, and reputation is identified. Another research in [35] developed a secure reputation algorithm for flooding for detection of malicious node. By means of a record-based system, security is increased with hybrid network for misbehavior detection. In [36,37], reference-based and table-based strategies are applied for behavior node detection with elimination of malicious node. In [6], tit-for-tat (TFT) strategy is developed for data forwarding based on incentive-aware routing. Research conducted in [14,18] utilizes credit-based and practical incentive schemes for message transfer in WSNs.

In [13], an active defense model is presented for active and dynamic management of defense strategy for file transfer. To improve efficiency of sensor network, [25] proposed an incentive mechanism for social network. In [23,38] competitive data forwarding and routing through hops is performed. For two-hop routing, asymmetric evolutionary game theory is developed in [39], with application of evolutionary

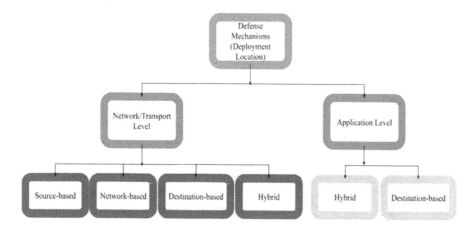

FIGURE 12.2 Classification of defense mechanism.

stable strategy (ESS). In Table 12.4, overall game theory based routing strategy and its advantage and limitation are presented.

12.4 OPEN ISSUES AND CHALLENGES

Even though several defense mechanisms have been introduced to handle attacks, the development and emerging technologies and network and platforms in this network, new threats and attacks are emerging every day, providing more challenges. Attacks are detected and estimated precisely, and defense operations are employed. In this section, ideal methods and existing network challenges and future works related to transport and application layers are taken into account to handle attacks [40–43].

By examining various defense methods in networks as well as application layers, it is seen that in several methods, work is centralized.

1. Since attacks are very complex and extensive, defense and detection methods need to work together.
2. No coherent with traffic protection is available in WSNs; thus, traffic protection has to be taken into account.
3. Defense mechanisms must be adaptive and possibly detects attack and non-attack patterns.
4. Because of restricted services given by WSN, requests are limited, so that legal requests are quickly detected and illegal requests are blocked from serving legal requests. WSN contains numerous malwares, so that needs to be estimated.
5. In WSN, security gaps and attack objectives are modified and comprehensive defense method with huge defense capacity was included. WSN devices are easily attacked, because they have fewer security features.
6. To enhance device vulnerability, device settings and security configuration in WSN are investigated to resolve existing gaps.

TABLE 12.4
Summary of Literature

Mechanism	Idea	Type	Advantages	Limitations
Sayed et al. [30]	Based on heuristic rules, compact and filtering, flow imbalance was detected.	Detecting and preventing	• Higher accuracy • Need of less computational resource	• In heavy traffic, it is inefficiency.
Wang et al. [33]	Detecting attacks based on examination of traffic for malicious	Detection	• Detection of low attack rate. • For responding, extract malicious address.	• For storage, memory is needed.
Gil and Poletto [34]	For collecting and detecting attacks, data structures are used in router.	Detecting and preventing	• In case of extensive attacks, legitimate traffic flow was stabilized.	• Random address failure. • Examined IPv4, for data packet transmission. • Storage shortage.
Tao and Yu [35]	Detecting attacks with the use of routers flow entropy.	Detecting	• During attack, exporting alert. • Real-time simulation.	• Based on assumption, restrictive is examined.
Ginde et al. [25]	Traffic monitoring	Monitoring and detecting	• Detection • Overhead reduction • Detection of attack.	• Increased computational cost • Increased Storage • Restrictive
Zhang et al. [14]	Attack was detected and prevented IP traffic.	Detecting and preventing	• Detection of flooding. • Reduced computational overhead. • Preventing attacks.	• To maintain data and storage consumption.
Wu et al. [36]	Detecting attacks by matching traffic patterns for decision tree	Identifying and detecting	• Using appropriate with estimation of positive and negative rate. • Attack detecting accuracy. • In short time intervals, it has ability to respond. • Between components it has secure communication channel	• Promising error in differentiation of traffic and attack detection. • Increased storage. • Increased complexity.
Thapngam et al. [37]	Based on estimation of traffic and pattern. identification.	Detecting	• Minimal false positive and false negative. • Applied in various detection like low traffic volume. • Improved detection rate	• Computational complexity. • Discriminating by the rate of packet transmission. • Detecting victim side wastes bandwidth of path. • Due to data review and observation, delay occurs.

WSN challenges are taken as future studies. Defense mechanisms are implemented by considering factors such as energy consumption, use of network bandwidth, and frequency as well as type of request made along with ratio of the data expected at any particular instance, location of data generated, resemblance of the traffic generated, and detection of adversary nodes.

12.5   CONCLUSION

This chapter presented a review of attack detection through a defense strategy-based mechanism. Further, defense methods are grouped as per the different layers, such as network layer and application layer. Application layer is divided into two classes, namely hybrid and destination-based method. For each class, this chapter reviewed and compared numerous methods. In distinguishing attack and legitimate traffic, the source-based method has no capacity in the transport/network layer. In case of failure in router or section, the network-based method fails. In rate limiting and traffic filtering, destination-based methods are not significant. Future work consists of implementing methods cooperatively by providing its corresponding platform. Application layer infrastructure must be strengthened, and effective collaboration between server and customer must be ensured, in order to conduct considerable protection against attack.

REFERENCES

1. Kumar B. and Bhuyan B., "Game theoretical defense mechanism against reputation based Sybil attacks", *Procedia Computer Science*, vol. 167, pp. 2465–2477, 2020.
2. Balachandran N. and Sanyal S., "A review of techniques to mitigate Sybil attacks", *arXiv preprint arXiv: 1207.2617*, 2012.
3. Vasudeva A. and Sood M., "Survey on Sybil attack defense mechanisms in wireless ad hoc networks", *Journal of Network and Computer Applications*, vol. 120, pp. 78–118, 2018.
4. Bhise A. M. and Kamble S. D., "Detection and mitigation of Sybil attack in peer-to-peer network", *International Journal of Computer Network and Information Security*, vol. 8, no. 9, 2016.
5. Newsome J., Shi E., Song D. and Perrig A., *"The Sybil attack in sensor networks: Analysis and defenses"*, *Third International Symposium on Information Processing in Sensor Networks, IPSN 2004*, pp. 259–268, 2004.
6. Sengupta D. K., Vakilinia I., Shetty S., Sengupta S., Kamhoua C. A., Njilla L. and Kwiat K., *"Three layer game theoretic decision framework for cyber-investment and cyber-insurance"*, *International Conference on Decision and Game Theory for Security*, pp. 519–532, 2017.
7. Tsuo F. Y., Lee W. L., Wang C. Y. and Wei H. Y., *"Power control game with SINR-pricing in variable-demand wireless data networks"*, *IEEE 71st Vehicular Technology Conference*, pp. 1–5, 2010.
8. Tushar W., Saad W., Poor H. V. and Smith D. B., "Economics of electric vehicle charging: A game theoretic approach", *IEEE Transactions on Smart Grid*, vol. 3, no. 4, pp. 1767–1778, 2012.
9. Byun S. S., Balasingham I. and Liang X., *"Dynamic spectrum allocation in wireless cognitive sensor networks: Improving fairness and energy efficiency"*, *IEEE 68th Vehicular Technology Conference*, pp. 1–5, 2008.

10. Chai R., Savvaris A., Tsourdos A. and Chai S., "Multi-objective trajectory optimization of space manoeuvre vehicle using adaptive differential evolution and modified game theory", *Acta Astronautica*, vol. 136, pp. 273–280, 2017.

11. Hakim K., Jayaweera S. K., El-Howayek G. and Mosquera C., "Efficient dynamic spectrum sharing in cognitive radio networks: Centralized dynamic spectrum leasing (C-DSL)", *IEEE Transactions on Wireless Communications*, vol. 9, no. 9, pp. 2956–2967, 2010.

12. Hao G., Wei Q. L. and Yan H., "A game theoretical model of DEA efficiency", *Journal of the Operational Research Society*, vol. 51, no. 11, pp. 1319–1329, 2000.

13. Luo Y., Szidarovszky F., Al-Nashif Y. and Hariri S., "Game theory based network security", *Executive Editor in Chief*, vol. 1, pp. 41–44, 2010.

14. Zhang Y., Wang J. and Liu Y., "Game theory based real-time multi-objective flexible job shop scheduling considering environmental impact", *Journal of Cleaner Production*, vol. 167, pp. 665–679, 2017.

15. Jiang C., Chen Y., Yang Y. H., Wang C. Y. and Liu K. R., "Dynamic Chinese restaurant game: Theory and application to cognitive radio networks", *IEEE Transactions on Wireless Communications*, vol. 13, no. 4, pp. 1960–1973, 2014.

16. Zhao Y., Wang S., Cheng T. E., Yang X. and Huang Z., "Coordination of supply chains by option contracts: A cooperative game theory approach", *European Journal of Operational Research*, vol. 207, no. 2, pp. 668–675, 2010.

17. Zhu M. and Martínez S., "Distributed coverage games for energy-aware mobile sensor networks", *SIAM Journal on Control and Optimization*, vol. 51, no. 1, pp. 1–27, 2013.

18. Ren H. and Meng M. Q. H., "Game-theoretic modeling of joint topology control and power scheduling for wireless heterogeneous sensor networks", *IEEE Transactions on Automation Science and Engineering*, vol. 6, no. 4, pp. 610–625, 2009.

19. Cittern D. and Edalat A., "*Reinforcement learning for Nash equilibrium generation*", *Proceedings of the International Conference on Autonomous Agents and Multiagent Systems*, pp. 1727–1728, 2015.

20. Abid I. B. and Boudriga N., "*Game theory for misbehaving detection in wireless sensor networks*", *The International Conference on Information Networking (ICOIN)*, pp. 60–65, 2013.

21. Bharathi M. A. and Kumar B. V., "Reverse game theory approach for aggregator nodes selection with ant colony optimization based routing in wireless sensor network", *International Journal of Computer Science Issues (IJCSI)*, vol. 9, no. 6, p. 292, 2012.

22. Wei X, Qu H. and Ma E., "Decisive mechanism of organizational citizenship behavior in the hotel industry – an application of economic game theory", *International Journal of Hospitality Management*, vol. 31, no. 4, pp. 1244–1253, 2012.

23. Wu Y. C., Tseng H. R., Yang W. and Jan R. H., "DDoS detection and traceback with decision tree and grey relational analysis", *International Journal of Ad Hoc and Ubiquitous Computing*, vol. 7, no. 2, pp. 121–136, 2011.

24. Dai T. and Qiao W., "Trading wind power in a competitive electricity market using stochastic programing and game theory", *IEEE Transactions on Sustainable Energy*, vol. 4, no. 3, pp. 805–815, 2013.

25. Gil T. M. and Poletto M., "*MULTOPS: A data-structure for bandwidth attack detection*", *USENIX Security Symposium*, pp. 23–38, 2001.

26. Niyato D. and Hossain E., "Radio resource management games in wireless networks: an approach to bandwidth allocation and admission control for polling service in IEEE 802.16 Radio Resource Management and Protocol Engineering for IEEE 802.16.", *IEEE Wireless Communications*, vol. 14, no. 1, pp. 27–35, 2007.

27. Koutrouli E. and Tsalgatidou A., "Taxonomy of attacks and defense mechanisms in P2P reputation systems—Lessons for reputation system designers", *Computer Science Review*, vol. 6, no. 2, pp. 47–70, 2012.

28. Fallah M., "A puzzle-based defense strategy against flooding attacks using game theory", *IEEE Transactions on Dependable and Secure Computing*, vol. 7, no. 1, pp. 5–19, 2008.

29. Raghava L. D., Niharika S. and Gupta A. S. A. L. G. G., (2021) Multi-objective evolutionary algorithms for data mining: A survey. In: Satapathy S., Bhateja V., Janakiramaiah B. and Chen Y. W. (eds) *Intelligent System Design: Advances in Intelligent Systems and Computing*, vol. 1171. Springer, Singapore.

30. Anand D., Rama Krishna Srinivas G., Gopala Gupta A. S. A. L. G., (2020) Fingerprint identification and matching. In: Satapathy S., Raju K., Shyamala K., Krishna D. and Favorskaya M. (eds) *Advances in Decision Sciences, Image Processing, Security and Computer Vision. ICETE 2019. Learning and Analytics in Intelligent Systems*, vol. 3. Springer, Cham.

31. Gupta A. S. A. L. G. G., Prasad G. S. and Nayak S. R., (2019) A new and secure intrusion detecting system for detection of anomalies within the big data. In: Das H., Barik R., Dubey H. and Roy D. (eds) *Cloud Computing for Geospatial Big Data Analytics: Studies in Big Data*, vol. 49. Springer, Cham.

32. Syamala M., Maguluri L. P., Gopala Gupta A. S. A. L. G., Akhila T. and Bhargav T. (2018) Optimized image compression method for portable devices. In: Sa P., Bakshi S., Hatzilygeroudis I. and Sahoo M. (eds) *Recent Findings in Intelligent Computing Techniques. Advances in Intelligent Systems and Computing*, vol 709. Springer, Singapore.

33. Bonnet G, "*A protocol based on a game-theoretic dilemma to prevent malicious coalitions in reputation systems,*" *European Conference on Artificial Intelligence*, pp. 187–192, 2012.

34. Sayed R. M., Khidr A. A. A. E. and Moustafa H. Z., "Changes in defense mechanism related to controlling Spodopteralittoralis larvae by gamma irradiated Steinernemacarpocapsae BA2", *Journal of Entomological Research*, vol. 39, no. 4, pp. 287–292, 2015.

35. Ho, J. W., Wright, M. and Das, S. K., "Zone Trust: Fast zone-based node compromise detection and revocation in wireless sensor networks using sequential hypothesis testing", *IEEE Transactions on Dependable and Secure Computing*, vol. 9, no. 4, pp. 494–511, 2011.

36. Abdelsayed S., Glimsholt D., Leckie C., Ryan S. and Shami S., "An efficient filter for denial-of-service bandwidth attacks", *IEEE Global Telecommunications Conference*, vol. 3, pp. 1353–1357, 2003.

37. Wang F., Wang H., Wang X. and Su J., "A new multistage approach to detect subtle DDoS attacks", *Mathematical and Computer Modelling*, vol. 55, pp. 198–213, 2012.

38. Tao Y. and Yu S., "*DDoS attack detection at local area networks using information theoretical metrics*", *12th IEEE International Conference on Trust, Security and Privacy in Computing and Communications*, pp. 233–240, 2013.

39. Thapngam T., Yu S., Zhou W. and Makki S. K., "Distributed Denial of Service (DDoS) detection by traffic pattern analysis", *Peer-to-Peer Networking and Applications*, vol. 7, no. 4, pp. 346–358, 2014.

40. Ginde S. V., Mac Kenzie A. B., Buehrer R. M. and Komali R. S., "A game-theoretic analysis of link adaptation in cellular radio networks", *IEEE Transactions on Vehicular Technology*, vol. 57, no. 5, pp. 3108–3120, 2008.

41. Younis M., Senturk I. F., Akkaya K., Lee S. and Senel F., "Topology management techniques for tolerating node failures in wireless sensor networks: A survey", *Computer Networks*, vol. 58, pp. 254–283, 2014.

42. Pandana C., Han Z., and Liu K. R., "Cooperation enforcement and learning for optimizing packet forwarding in autonomous wireless networks", *IEEE Transactions on Wireless Communications*, vol. 7, no. 8, pp. 3150–3163, 2008.

43. Abrardo A., Balucanti L. and Mecocci A., "A game theory distributed approach for energy optimization in WSNs", *ACM Transactions on Sensor Networks (TOSN)*, vol. 9, no. 4, pp. 1–22, 2013.

13 Securing Wireless Multimedia Objects Through Machine Learning Techniques in Wireless Sensor Networks

Rakesh Ahuja, Ambuj Kumar Agarwal, and Manish Sharma

Chitkara University, Punjab, India

Purnima

Sri Guru Gobind Singh College, Punjab University, Chandigarh, India

Saira Bano

Vel Tech University, Chennai, India

CONTENTS

DOI: 10.1201/9781003145028-13

13.1 INTRODUCTION: WIRELESS NETWORK

A computer network is considered a wireless networks if no cables of any type are used to connect to two or more computers. The necessity of wireless networks arises to avoid the costly and complex process of establishing the connection through cable structure. The wireless system is based on radio waves, which use the physical layer of network to implement the networking of electronics devices or Gazettes as laptop, palmtop, iPad, TV mobile, etc. In general, there are four major categories of wireless networks: wireless local area network, used to link electronics devices by wireless distribution methods; wireless metropolitan area network, which links numerous wireless LANs; wireless wide area network, to cover huge locational geographic area as cities and towns; and wireless personal area network, to interconnect devices for a short span of time.

13.1.1 WIRELESS SENSOR NETWORK

Wireless sensor networks (WSNs) can be well delineated as self-built wireless networks that do not have an infrastructure to supervise physical or ecological circumstances, such as high surface temperature, vibration, density, movement, or contaminants, and for binding information to pass through from the system to the main site (sink). The sink acts as a transit point between the wireless network [1] and the users. The user can reclaim mandatory data from the network by entering queries and collecting the consequences from the sink node. The wireless sensor network covers millions of billions of device nodes. The device nodes can be interconnected using radio signals. Wireless communication is increasing at a rapid rate and will play a vital role in network access. This can be seen through the extensive implementation of WLAN and cellular networks and the advent of cognitive radio networks. These networks, measured as wireless access, are generally connected through a basic wired network. A strengthened support can host a server, or can be associated to a quick wired router, that connects to the cyberspace. The structural design of wireless communication [2] is shown in Figure 13.1.

The wireless sensor nodes are efficient with lights, computers, wireless transceivers, and power routes. One node in the WSN has many restricted resources. They have restricted production bandwidth, speed, and storage capacity. Once one of the sensors is classified, they are responsible for self-regulation of a suitable network structure and common and multi-hop communication with them. The radio sensor also responds to information transmitted from the "control site" to execute precise instructions or to connect parts of the sensor. The sensor node activation path cannot be interrupted, but it is driven by event. Global Positioning System (GPS) and classification procedures

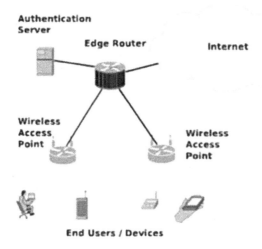

FIGURE 13.1 Architecture of wireless communication network.

are employed to locate and record information. The actuators equipped in wireless sensors operate under special conditions. These connections are sometimes explicitly referred to as the wireless sensor network and the network interface.

WSNs allow innovative applications and integrate unusual design patterns due to the limited size. Given the limitations of the small complexity of the method and the low power consumption (long speed system), the right balance should be struck between signal and potential action information. This has been a major factor in research activities, modeling methods, and industries invested in this area since recent times. Nowadays, many WSN network researchers have introduced powerful and computationally efficient blockchain models and algorithms, and the application component is designed for easy information-oriented communication and monitoring software.

13.1.2 OBJECTIVES OF WIRELESS SENSOR NETWORK (WSN)

Chief objectives for sensor positioning are identified as network connectivity and longevity, area coverage, and data conformity. Here, we provide a brief outline of its objectives.

13.1.2.1 Coverage

Coverage/reach is a largely measured goal. The dilemmas of coverage are point/target coverage, region coverage, power proficient and k-coverage issue. The coverage estimate differs based on the fundamental field of individual sensor and the metric used to calculate the cumulative coverage of installed sensors. The utmost familiar sensor cover model is the sensor of disc model. It is measured that all objects in the sensor-centered disc are covered by the sensor. The detection capabilities in the sensor's coverage area can be divided into zero or one coverage model, also known as the binary model, information coverage model, and probabilistic coverage model.

13.1.2.2 Differentiated Detection Levels

Deploying differentiated networks of sensor which reproduces contentment with detection rates at altered geographic locations is also a significant subject. Nevertheless, for those certain parts that are not as sensitive, relatively low probabilities of detection are mandatory in order to diminish the number of sensors deployed to reduce costs. So the changed vicinity necessitates diverse densities of expanded nodes. Consequently, detection needs are unevenly disseminated in the vicinity. As a consequence, the approach of the WSN organization must take into account the geographic features of the observed events.

13.1.2.3 Network Connectivity

Another challenge when designing a WSN is network connectivity. We say that a network is connected if any vigorous node can converse with any other vigorous node (possibly using other nodes as relays). A network connection is obligatory to ensure that messages are transmitted to a suitable sink node, as well as to interrupt communications if this often persists at the end of the network's life. It has a lot to do with coverage and energy efficiency. The relationship between coverage and communication results from the detection and transmission range. If a node's transmission range extensively exceeds its detection range, then connectivity is irrelevant, as coverage ensures that there is a way to communicate. The conditions are poles apart if the communication range is less than the operating range.

13.1.2.4 Network Life Span

The foremost confronts in developing a WSN is communicating that energy wealth is very restricted. Revitalizing or substituting the battery of a sensor in a system can be complicated or unmanageable, resulting in rigorous precincts in communication and synchronization between all sensors in the system. Note that the breakdown of the requested sensors may not damage the overall functioning as adjoining sensors can take over if their density is soaring. Consequently, the chief constraint to be improved is the life span of the network, the period until the network is partitioned in such a way that collecting data from one part of the network becomes unbearable.

13.1.2.5 Data Fidelity

Validating the trustworthiness of the accumulated information is a significant aspiration of the WSN project. The sensor system provides a combined calculation of the perceived phenomena by combining the readings of several self-governing (and sometimes different) sensors. Combining data improves the trustworthiness of reported events by plummeting the probability of false positives and the disappearance of the detected object. Increasing the number of sensors reporting in a given region will definitely perk up the accuracy of the aggregated data. Nonetheless, termination in coverage needs a higher node density, which may be unfavorable due to an increase in cost or a decrease in survivability (the possibility of detecting sensors on the battlefield) [3].

13.1.2.6 Energy Efficiency

This principle is often used as a synonym for service life. Due to the restricted energy source in every sensor node, we have to apply the sensors in a well-organized way to boost the period of the network. There are at least two ways to solve the predicament of power preservation in sensor networks associated with the ideal location. The first method is to program active sensors, which allows other sensors to enter catnap mode by applying overlaps between measurement ranges. The second method is to change the response range of the sensors to maintain power.

13.1.2.7 Imperfection Tolerance and Load Balancing

Imperfection-tolerant design is imperative to evade isolated faults due to limited network life. Many authors focus on the formation of k-linked WSNs. The k-connectivity assumes that there are k self-governing paths between each pair of nodes. For k ≥ 2, the system can accept some node and channel breakdown. Due to the many-to-one communication scheme, the k connection is a particularly significant design factor in the base station area and promises some bandwidth for communication between nodes.

13.1.3 WSN Relevance

WSNs have grown in fame because of their ability to solve hitches in a variety of applications and are likely to change our lives in numerous ways. WSNs are productively applied in several areas as follows.

13.1.3.1 Armed Forces Applications

WSNs can be a fundamental part of armed forces command, control, communications, computing, reconnaissance, surveillance of battlefield, scouting, and guidance systems.

In area monitoring, sensor nodes are arranged in the area where an incident is to be observed. When the sensors detect an observed event (pressure, heat etc.), the event is transmitted to one of the sinks.

Instantaneous traffic statistics are compiled from the WSN to utilize more transportation models and alert drivers to traffic and overcrowding issues.

Healthcare applications for sensory systems are assistance interfaces for people with disabilities, integrated patient supervision, diagnosis and supervision of drugs in sanatorium, remote supervision of individual physiological data, and follow-up and supervision of doctors or patients at the sanatorium."

The term "Environmental Sensor Networks" was created to encompass a lot of the WSN requests for Earth discipline study. This comprises forests, oceans, volcanoes, glaciers, etc. Other important studies are as follows:

- Greenhouse monitoring
- Air pollution supervision
- Detection of landslides
- Woodland fire recognition

Sensors can be utilized to supervise work on construction and setup, like flyovers, tunnels, overpasses, embankments, and more, allowing engineers to keep an eye on resources remotely without the actual visit of sites, which requires a huge cost.

WSNs were created for condition-based maintenance (CBM) because these networks provide noteworthy expenditure, enabling and saving novel functions. In cabled systems, the connection of adequate sensors is frequently restricted through the cabling charges.

The wireless system prevents the cultivators from storing wiring in complicated conditions. Irrigation mechanization enables additional proficient utilization of water and diminishes ravage.

13.1.4 WSN Features

A WSN is composed of numerous diverse components, of which the sensor assembly is a significant but tiny element. The characteristics of a valuable WSN embrace energy proficiency, scalability, performance, trustworthiness, and mobility. A WSN with such structures can be exceptionally precious and, if left unchecked or unsecured, can lead to network congestion, making it unsuitable.

13.1.4.1 Power Efficiency in Wireless Sensor Networks

Energy efficiency is permanently the main concern as the process of WSN hinges deeply on the lifecycle of the battery of sensor nodes. The activity to route the data packet is the utmost energy-consuming process in a WSN. Power efficiency is the

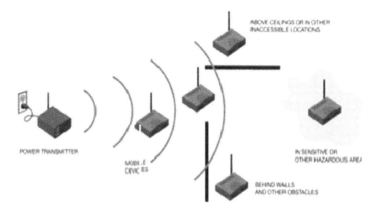

FIGURE 13.2   A wireless power distribution system (image courtesy Powercast).

system's capability to grip mobile nodes and variable data paths as shown in Figure 13.2. Moreover, it is the system's capability to grip mobile nodes and variable data paths. It is an exceedingly significant factor, particularly since it frequently occurs with WSNs that the sensor nodes are remotely placed without correct contact to a power point. The design of a wireless sensor network must be exceedingly sensitive to cope with mobility. As an upshot, it is becoming gradually more complicated to devise a comprehensive WSN with mobility. The device may devour less power to do more if it is operating at very stumpy power levels. Consequently, these procedures are typically built to operate from a power source other than undeviating electrical energy. The most excellent design method would be to diminish the duty cycle of each node [4]. A wireless power allocation system can endow with coverage for a full facility similar to a lighting system or wireless communication system.

13.1.4.2   WSN Scalability

The capability for a network to develop in terms of the number of nodes committed to the WSN without producing unnecessary fixed cost can be called its scalability. The simple implementation of such a network is composed of only a handful of nodes, and they must endow with support for more as well.

13.1.4.3   WSNs Responsiveness

The capability of a network to swiftly acclimatize to changes in topology is regarded as its speed. Nevertheless, a very accessible web has drawbacks; cooperation is necessary. The packet-forwarding potential in a dynamic environment as well as scalability will diminish the enormously accessible network. The sensitivity structure of a WSN is revealed in Figure 13.3.

13.1.4.4   Steadfastness in Wireless Sensor Networks

Any network wants to be unswerving, that would be a straightforward precondition. They need reliable data transmission in the ever-changing state of the network composition. In general, there is a converse association between scalability and

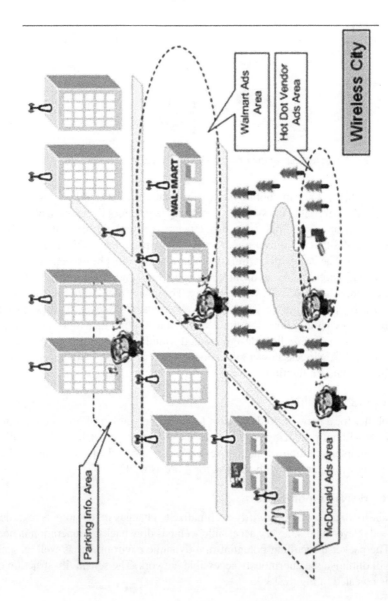

FIGURE 13.3   Wireless sensor network responsiveness (image credit: multimedia.ece.uic.edu).

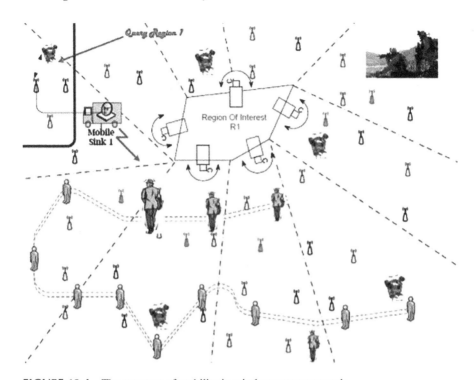

FIGURE 13.4   The structure of mobility in wireless sensor network.

dependability in WSNs. As the number of nodes in WSN enhances, it turn out to be more complicated to preserve steadfastness. If the network is tremendously scalable and scales to a larger network than firstly designed, this will put pressure on the reliability of the data transfer, and the breakpoint will appear earlier.

13.1.4.5   WSN Mobility

Mobility is the capability of the network to tackle mobile nodes and impulsive data paths. The main design requirement is that the WSN is extremely sensitive to cope with the mobility. As a consequence, it turns out to be more complicated to develop a large-scale WSN. An exemplar of mobility in a WSN is shown in Figure 13.4.

13.1.5   SN Categories

Depending on the situation, categories of wireless networks are selected that require to be used underwater, on land, underground, etc. The diverse categories of WSN embrace the following.

13.1.5.1   Ground-Based WSNs

Ground-based WSNs are achieved through competent base station communication and are composed of millions and billions of nodes arranged in an amorphous or

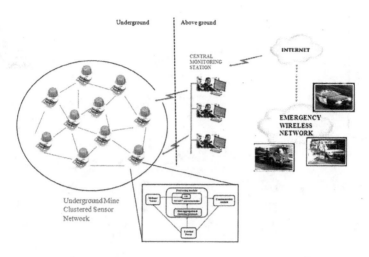

FIGURE 13.5   Underground WSNs.

prearranged manner. In amorphous mode, nodes haphazardly move in the target area, falling from an unchanging plane. The prearranged mode takes into account the optimal location, grid location, and two-dimensional as well as three-dimensional models. In this type of sensor network, the battery capacity is restricted; nonetheless, the battery is powered by secondary energy source like solar cells. Power preservation of these network is realized through the use of low duty cycle operations, diminishing latency, finest routing, etc. [5].

13.1.5.2   Underground-Based WSNs

Underground types of wireless sensor networks are higher priced than ground-based WSNs in terms of implementation and maintenance. Such a network contains numerous sensor nodes that are secret on the ground for monitoring underground circumstances, as shown in Figure 13.5. There are additional receiver nodes above the ground to transfer statistics from the sensor nodes to the sink [5].

Underground WSNs located in the land are thorny to revitalize. Well-equipped sensor battery assemblies with restricted battery life are complicated to recharge. Besides this, the underground environment makes wireless communiqués knotty due to high signal dwindling and loss [5].

13.1.5.3   Underwater Based WSNs

More over 70% of the earth's surface is covered by water. These networks comprise numerous vehicles and sensor nodes located under water, as shown in Figure 13.6. Self-governing underwater vehicles are used to set the data of these sensor nodes. One of the challenges of subsea communication is long latency, throughput, and sensor failures. WSNs have a restricted capacity underwater battery that cannot be invigorated or substituted. The predicament of power preservation for subsea WSN is associated with the development of subsea networks and communication technologies [5].

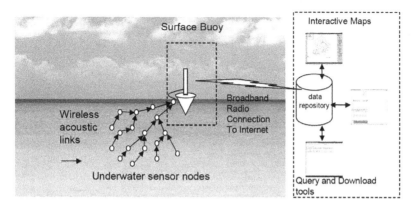

FIGURE 13.6   Underwater WSNs.

13.1.5.4   Multimedia WSNs (M-WSNs)

M-WSNs have been planned to track as well as observe events in the form of multimedia categorized as imagery, video, and auditory. It is composed of inexpensive sensor nodes stimulated with camera and microphones. In this kind of network, nodes are organized through a wireless connection to compress the data, recovery, and for association as shown in Figure 13.7.

13.1.6   Unauthorized Access Point Detection in WSNs

The malicious access point is categorized into two major parts, as described in the following.

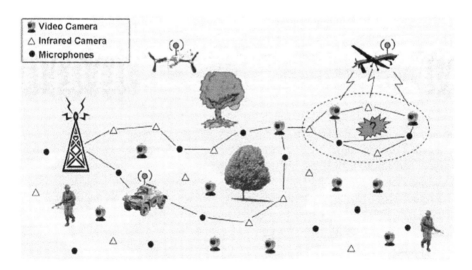

FIGURE 13.7   Multimedia WSNs.

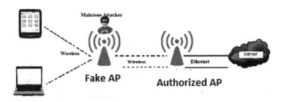

FIGURE 13.8   An example of fake AP attack. [3]

13.1.6.1   Fake Access Point

It is formed or connected by a malicious attacker which is not part of the network. Without approving users, it performs attacks like denial of service or a man-in-the-middle. It is set up by a malicious attacker for malicious behavior such as fabrication, overhearing, or stealing evidence [3,6].

13.1.6.2   Rogue Access Point

A rogue access point, also called a rogue AP, is any Wi-Fi access point that is connected to a network but is not approved to operate on that network and is not part of the network administrator's organization. The term "rogue access point" has been used in more than one context in the area of wireless security. It is mounted not only or configured by an external attacker, but also by an approved user on the network to further take advantage of the network. It is very unproblematic to create a fake access point, as shown in Figure 13.8. The attacker creates his access point using some commercially available software, as shown in Figure 13.8. After creating a fake access point, the attacker waits to ask the client node to connect to this rogue access point or passively send several signals to the client node from time to time and turn it on to change the connection. It even inspects wireless traffic with tools like the Aircrack-ng package, releases the beacon and control structure, and attempts to obtain the node MAC address, logical address, and service set identifier (SSID). Using this method, he manages to attack client nodes. Without creating an additional network connection, it uses internet services for the wired network through an approved access point and provides them to the client node. Thus, the attacker negotiates the individual evidence of the client node without knowing the client. The main goal is to improve the software by detecting and countering the rogue access point using Advanced Internet Proxy. This contributes to the new rogue access point discovery method in the network [2,4]. Rogue APs, if undetected, can become an open door for sensitive information on the network. Undetected rogue APs at enterprises were used by many raiders not only to gain free access to the internet, but also to view confidential information. Most of today's rogue AP discovery solutions are not automated and rely on a specific wireless technology.

Rogue Access Point Detection & Counter Attack is a desktop application that detects APs first and checks if this AP is a rogue AP. The detailed architecture is described in Figure 13.9, demonstrating that the proxy server plays an important role. This proxy will run on the server. There is another class that runs and receives all the

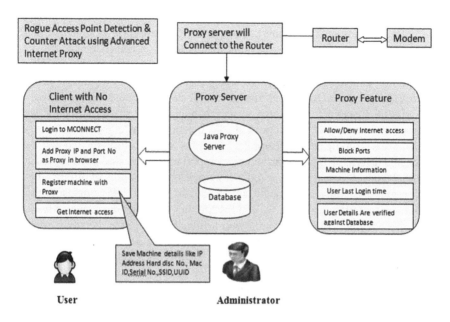

FIGURE 13.9   Architecture of Rouge access point detection & counter attack.

details from the machine. This data is then stored in the database. When checking to see if a request is approved, the proxy crossovers check the database and, based on the details, provide them with internet access. The system architecture is implemented as a client-server architecture using standard commercial equipment and an open source Wi-Fi tracker. A modified version was created and configured on multiple monitors as clients. These clients enhance AP data transmission and signal appearance. They then report their amounts to the server, which notices and locates rogue APs, if any [4].

13.1.7   WIRELESS MULTIMEDIA SENSOR NETWORKS

With the emergence of several technologies related to multimedia, operations with respect to wireless multimedia data become effortless and expedient. As the usage of multimedia has increased day by day, the chances of attack on wireless multimedia data also increase. In order to maintain the trustworthiness on the digital multimedia data, the same should be well protected [7]. Additionally, there are more chances of infringement in the copyright, authentication, and confidentiality of wireless multimedia data. The major area of concern is the storage and computation, and they need to be taken care of. An important challenge here is to be ready for handling diverse security threats as well as potential attackers. At the present time, artificial intelligence has been extensively used in almost all the fields. The major branch of artificial intelligence is machine learning that is widely used in the area of image processing, natural language processing, pattern recognition, recommendation-based system, healthcare, prediction of stock market, etc. The most important challenge at present is to manage the security of wireless multimedia data like audio, image, video, and motion pictures by using different and novel information security techniques. Furthermore, with the evolution of new techniques in the field of wireless multimedia security, there is a need to find solutions to problems that occur while exploring such techniques, for example, to enforce secure channel for wireless multimedia data to safeguard the privacy of the same with the functioning of cloud computing. The motive of this write-up is to concentrate on various security issues related to wireless multimedia data implemented through the support of the machine learning process.

With the development of advanced technologies, the networks of multimedia and their progressive union are broadly anticipated to fetch exhilarating facilities for scrutinizing, entertaining, guiding, and functioning in the field of smart city, smart home, medicine and healthcare, etc., with artificial intelligence. People's everyday lives have completely changed and become much more effortless with the use of several multimedia applications [8], although the advent of technology and advancements made the creation of these multimedia systems complex and also carries numerous protection issues. Taking an example, if a multimedia channel gathers a particular amount of information with the help of sensors, then few malevolent sensors could always mislead the user by offering counterfeits. Numerous latent threats could direct to some terrible results and make severe damage in multimedia networks using wireless mode.

Furthermore, the most common components used for multimedia are prone to risks which will in turn originate prevalent threats to the network through diverse modes like spoofing and eavesdropping. Thus, there exists a need to develop an efficient and successful safety system for wireless multimedia communication to guarantee the safety of the data transmission. The verification as well as authentication methods are generally used as the base mechanism and significant plan for keeping the multimedia paradigm secure and safe from all aspects as the opponent attacker requires the information to access the resources [9]. All the schemes help the multimedia network to protect the genuine information by verifying the users' identity and giving them access to the legitimate network. In case of a greater number of multimedia components connected over a network, the development of a secure channel for the communication pretenses a great challenge. But the multimedia machines alleging fewer holdup transmissions could not sustain the validation approaches, as these need elevated transparency of calculations and incredible physical components (sensors) present in the wireless multimedia structure, involving little computational outlay to certify the performance with respect to communication. Therefore, to shield the wireless multimedia network against the risk of attackers, our chapter focuses on the performance analysis as well as the challenges occurred by conventional methods of authentication and also recommends novel security improvement schemes (Table 13.1).

13.1.8 LITERATURE SURVEY

TABLE 13.1
Literature Survey

Ref. No.	Algorithm/Application	Outline
[10]	A sorting algorithm for maintaining privacy based on the logistic maps for clouds and arranging the data efficiently.	The outcome of the experiment and the design of the technique suggested that the privacy of the data was well protected and also offered the cataloging of the coded data in a proficient manner.
[11]	The information reversible to hiding system based on CDMA technique and methods comes under machine learning. It handles the issues of medical imaging.	As per the results of the research, the method proposed here was able to attain the exceptional figures with respect to insertion capability of the data as well as to deal with the security issues occur in the field of medical imaging.
[12]	A re-encryption method of multimedia data based on conditional proxy re-encryption key addresses the problems of wireless communication.	The paper deals with the challenges faced during the operation of wireless communication and was carried by exhibiting nested encryption on the wireless multimedia data by using a broadcast re-encryption key.
[13]	An image searching system based on deep learning hash algorithm and to identifying the malicious objects and images.	The paper demonstrated a method which was executed in different subparts. These subparts involved the feature extraction process, deep learning image search as well as the classification of the image. The results of the said approach suggested that the system was able to identify the unlawful images.
[14]	Elaborated a system for identifying the wireless device based on ensemble learning.	The experiment here was composed of three segments, including extraction of RFF, detecting signal, and the classification model. The consequence showed that this identification technique was much more reliable and produced much better results than the other methods available in the respective field.
[15]	An intrusion detection system based on feature selection algorithm.	The said approach used self-adaptive differential evolution method for intrusion detection, and the outcome of the procedure showed potential results as compared to other methods.
[16]	Permission Management System based on the machine learning algorithm protecting and securing the Android applications.	Machine learning is used here to develop an enhanced and active permission management system in order to safeguard the Android applications. The outcome of the experiment is that the suggested technique improves the areas of relevance of the permission management.
[17]	Mixed Linear Integer Programming Approach uses lightweight block ciphers to address safety issues of the ciphers	In this paper, the authors addressed the security issues related to lightweight block ciphers by suggesting a mixed programming method. The proposed scheme can decrease the quantity of the variables extensively, maintaining the efficiency of the system simultaneously.
[18]	Single image de-raining algorithm based on generative adversarial networks enhances the quality of the image.	In order to augment the image quality, the algorithm was designed that provides an outstanding output, keeping in mind the speed of computation with the help of generative networks.

(Continued)

TABLE 13.1 (Continued)
Literature Survey

Ref. No.	Algorithm/Application	Outline
[19]	A switching technique considered as slot-wise is proposed. This is for optimization of bootstrapping method.	The main aim of this switching approach was to optimize the bootstrapping method as well as to secure the multimedia data present in the cloud. The result showed the reduction in computing speed and storage intricacies.
[20]	Detecting fraud and protecting security issues while doing transactions online. The technique they use is the convolutional neural network.	To secure the online transactions done by the users, a fraud detection system was proposed in this article which was implemented using the convolutional neural network. As the input data was low dimensional, the results were quite impressive as compared to other, similar methods.
[21]	Proposed a system to detect the interference. It is based entirely on payload and statistical features.	This scheme was given to retrieve the valuable information from the data payload with the help of a text-convolutional neural network and finally applying the random forest algorithm on the approach to get the best results to preserve the sheltering of the multimedia information.

13.1.9 PARADIGMS OF INTELLIGENT AUTHENTICATION FOR EFFICIENT MULTIMEDIA SECURITY

The major basis of algorithms designed under machine learning can be broadly characterized by two specific features i.e., working and their arrangement giving away the following categories of algorithms as decision tree algorithms, clustering algorithms, regression algorithms, and Bayesian algorithms.

In this chapter, we elucidate various techniques of machine learning from two different viewpoints—parametric as well as nonparametric machine learning schemes; other techniques are unsupervised, supervised, and reinforcement methods [9,10]. This categorization of machine learning methods is shown in Figure 13.10. The terms

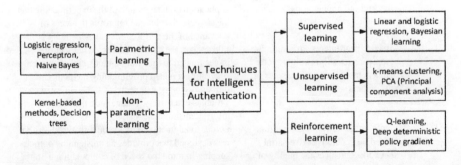

FIGURE 13.10 ML techniques for intelligent authentication.

used to denote these categories, like parametric and nonparametric, point towards whether there are precise forms of guidance functions; supervised and unsupervised designates whether the samples are characterized in the storage; and reinforcement learning indicates if there exists equilibrium between investigations of unexplored area and utilization of progressive understanding. Here, we have presented the fundamental idea of various machine learning techniques and also talked about their areas of relevance and the prerequisite in creating the authentication system for diverse wireless communication situations [22].

13.1.9.1 Parametric Learning Methods

Parametric learning methods are very important in literature and are shown by logistic regression and the Naive Bayes methodology since they require special features of guiding. The outcomes of the parametric approaches may be more accurate and easier, and fewer samples are required than nonparametric approaches, as the proper guiding functions are selected. The algorithms, based on parametric techniques of learning, typically comprise two steps: choosing a contour of the function, and investigating the function coefficients from the guidance information. In case of intelligent authentication, parametric methods could represent the elements separately grounded on the explicit form of training methods so that the ambiguities may be avoided by the composite time-varying situation.

13.1.9.2 Nonparametric Learning Methods

On the other hand, nonparametric methods are not that specific, but they are extracted from the presented information. The most popular examples of these methods are k-nearest neighbors and decision trees. Nonparametric methods become skilled vigorously from time-varying surroundings by not involving any kind of assumptions regarding the guidance model. Constructively, these methods provide elevated flexibility for the authentication system as they are competent of fitting a heavy count of purposeful forms. Thus, the outcome of such methods would be of an advanced level for forecasting. However, these approaches need a lot more training data to approximate the mapping function, and they have a poor training speed since they often need a lot more parameters to direct them. Due to this, there exists greater risk to overfit the data, and sometimes, it may raise questions about the exact prediction which has been prepared.

13.1.9.3 Supervised Learning Algorithms

This is the first kind of machine learning approach whose job is to study the mapping role with the labeled data set [23]. Here, the scheme is precisely instructed to predict the exact category of the new data object. Therefore, the prototype is guided to perceive the essential patterns as well as connections which assists it in producing high-quality and accurate results. In particular, supervised learning can be categorized into two basic types, as follows:

a. *Regression*: Regression algorithm tasks are required to manage the data in a continuous form. In the regression technique, a best fit line needs to be found to get more precise output. The regression model is an efficient method to predict the price for real estate, or to predict click-through rates for online advertisements, or to predict at what price a customer would buy a particular product.

b. *Classification*: A classification algorithm regulates the category or class of the data objects. In other words, the process of distinguishing, considering, and making the cluster is known as classification. For instance, categorizing incoming emails under "spam" or "non-spam" is the best example of using a classification algorithm.

These categories are based on the aspect of the stability of the output. Many types of algorithms exist to classify data, such as neural networks, decision trees, logistic regression and naive Bayes, stochastic gradient descent, decision trees, random forest, and k-nearest neighbors. Out of these, some of the popular algorithms are described next.

Support vector machine (SVM): It is a type of linear classification scheme that separates various data points to find out the best plane which possesses great margins between two given section of data sets, and it also supports multiclass [24]. This is also known as maximum margin plane. Generally, the data set is not that easily divisible, but in order to do that, the actual space can be plotted into a superior dimensional planetary system to a great extent by using kernel functions like polynomial or Gaussian functions which are popular for nonlinear classification.

K-nearest neighbors: It is a nonparametric learning scheme used most generally in the case where no supposition of the distribution-related area is required like classification and regression. The basis of this technique is to settle on the section of the data set where feature vector is based on maximum in demand k-nearest neighbors. The disadvantage of this method is to separate the section of the data set used frequently as it sometimes governs prediction results. For this, a weighted approach can be implemented on the same to use the weight for an individual neighbor that is related to the distance between the sections in the opposite direction.

The major dissimilarity between the two learning methods, known as supervised and unsupervised, is that the supervised learning needs prior information about the inputs used and resultant outputs. Conversely, unsupervised learning is the one which does not involve the usage of labeled data sets. Few supervised learning algorithms as per the literature has been discussed in the article [25] for diverse application areas with the usage of SVM and k-nearest neighbors. While choosing the relevant algorithm for implementation of a specific problem, one should keep in mind the size of the data set and the type of authentication problem; mostly labeled outputs are required for real-time authentication problems.

13.1.10 UNSUPERVISED LEARNING ALGORITHMS

This type of learning is a guidance program for the machine to use information which is neither classified nor labeled and to have the machine operate on that information without any training. It can also be explained as the machine learning job whose motive is to study a particular utility to depict a concealed arrangement. Usually, the given unsupervised learning algorithms are used in surveys. In this method, the mission of the machine is to merge the unsorted information as per the likeness and disparity, with no appropriate assistance of data. In contrast to supervised learning, no trainer is allotted, and therefore, no training is specified to the machine. Hence, the machine has to determine the information and patterns that were initially unobserved. Unsupervised learning is divided into two sets of algorithms:

a. *Clustering*: It is the category where the user wants to notice the intrinsic groups in the information; for example, the customer's behavior according to their way of purchasing the groceries.

b. *Association*: In this problem, the user wants to notice the imperative that determines the large part of the information. An example is determining that customers who buy x are inclined to buy y.

The various categories of methods for clustering are agglomerative, probabilistic, exclusive, and overlapping. Moreover, these are classified as hierarchical clustering, principal component analysis, independent component analysis, singular value decomposition (SVD), k-nearest neighbors, and k-means clustering. The feature of abnormality detection can find out significant points in the data set that helps in locating falsified transactions, but then too, it cannot provide the accurate information about the data.

13.1.10.1 Reinforcement Learning Algorithms

The main objective of this learning is to observe the ideal behavior of the components, in turn, to maximize the efficiency and performance of the system. This method of learning is based completely on the experience, as it does not involve the labeled data set and the results related to them. With the execution of the system and the actions performed, the system itself judges and learns which all actions are required to perform to maintain the competence of the method. In the said scheme, the representative focuses on the optimization of the algorithm with the help of the surrounding based interactions. The reinforcement algorithms which are distinctively applied in several surveys do not need that level of precision in input and output parameters (Table 13.2).

13.2 IMPLEMENTATION OF MACHINE LEARNING ALGORITHMS IN MULTIMEDIA SECURITY

This section highlights the various machine learning-based implementation techniques to securely transfer the multimedia contents in an authentic way through a wireless network.

13.2.1 SUPERVISED LEARNING ALGORITHM

The resource allocation issues based on chance constrained energy efficiency have been studied for a multicast network [26] considered as OFDM. The main aim of [26] is to maximize the effectiveness of the complete system. To solve the problem related to power allocation, a function needs to be defined as per the probabilistic

TABLE 13.2

Performance Analysis of Machine Learning Authentication Techniques

Method	Details	Use	Drawback	Application in the Field of Security
Decision Tree	A prediction model is established by using DT, and training samples are represented as branches and leaves to study from them. Then, to predict the class of the new sample, the pre-trained model is used.	• Transparent method • Simple • Easy to use	• Because of its construction nature, it needs a lot of storage • It is easy to understand if a handful of DTs are associated	• Intrusion • Suspicious traffic sources
Support Vector Machine	They proposed the technique to maximize the distance between the hyper-plane and the greatest neighboring sample points of each category. SVM makes a splitting hyper-plane in the feature dimension of different classes.	These machines are best known for their oversimplification ability. The most appropriate part of data comprises a greater number of feature aspects and in contrast considers a smaller number of sample points.	• It is tough to select the optimal kernel • It is a challenging process to interpret and understand SVM-based models	• Intrusion • Malware • Attacks in smart grids
Naïve Bayes	In NB, posterior probability is calculated. It forecasts the probability that the unlabeled samples' feature set suits a specific label by assuming that the features are independent of each other.	• Simple • Easy to implement • Requires low training sample • Robustness to the unimportant features and preserve it independently	• It does not hold features independently, so through the relationships and interactions between the features, it cannot collect effective solutions	• Network intrusion
K-nearest neighbors	KNN organizes the new samples and decides its class through the votes of its nearest selected neighbors.	• Effective and well-liked method for intrusion detection	• The process of determining the optimal value of k is difficult and consumes lots of time and effort, as the optimal value of k generally varies among different data sets	• Intrusions • Anomalies

(Continued)

Method	Details	Use	Drawback	Application in the Field of Security
Random forest	To achieve refined and enhanced results, several DTs are constructed and joined to get the effective prediction model.	• The scheme is vigorous to overfitting • The attributes are selected through bypass methods for feature selection • Much fewer parameters are required for input	• It is not applicable in specific real-time applications where a large training data set is needed, as several DTs are constructed in it	• Intrusion • DDoS attacks • Unauthorized IoT devices
Augmented reality (AR) algorithm	In the AR algorithm, the correlations of variables are detected by studying the variable relationship in a given training data set. Then, accordingly, a model is constructed.	• It is used to specify the category of novel trials • Considered simple and easy to use	• Time complexity is high • It uses simple assumptions between variables, i.e. direct relationships and occurrence • It is not applicable to security applications	• Intrusion
K- means clustering	It is based on an unsupervised learning model. It defines clusters in the data on the basis of similar features by varying the value of K. It also signifies the quantity of clusters formed by the algorithm.	• Suitable for data considered to be private in an IoT scheme • Labeled data is not required	• The scheme is less capable those methods comes under supervised, precisely to detects identified outbreaks.	• Industrial WSNs to detect Sybil • IOT based system contains anonymous data considered to be private
Principal Component Analysis	It is a conversion process that converts the principle components, i.e. the amount of correlated parameters, into a lower numbers of parameters but must be uncorrelated.	• It reduces the complexity of the model by achieving dimensionality reduction	• It needs other ML methods to set up a powerful security approach, because it is a feature reduction method	• IoT environment provides a real-time system to detect and reduce the model's features

interference constraint which is calculated by the support vector machine algorithm of machine learning. This technique increases performance in chance-constrained settings while also increasing the joy index and energy effectiveness [27]. A power control and twofold ray forming method for the communication of inter devices has been given to address the challenges that occur when using them. The implementation has been done in the steps starting with the formulation of technique to reduce the total power transmitted from the devices associated in the network for maintaining quality of service and suppressing mutual interference conditions. An approximation method has been proposed using the support vector machine algorithm to solve the optimization problem for power transmission as well as for beam forming weight vectors related to the users [28]. A dynamic cluster-based, cost-effective mechanism for small cell base stations has been proposed, and the focus of the research was to enhance the complete performance by using the features like their competence to manage the traffic dynamically and the information on both positions of the base station. To implement the scheme, both centralized and decentralized clustering techniques are given and the outcome of the said technique was much better. The method also suggested the way to take the maximum use of the benefits of cluster-based schemes in small cell networks [29]. An integrated algorithm of machine learning, named the k-means clustering method, was used to get better results related to the allocation of the spectrum, balancing of load, and BS position and sleep mode action. The use of several machine learning algorithms provides an effectual balance between energy efficiency and quality of service, which further enhances the capacity of the network.

13.2.2 REINFORCEMENT LEARNING ALGORITHM

A distributed user-association approach is demonstrated by the use of algorithms based on machine learning to promote vehicular networks [30]. The major focus of this approach was on the enhancement in load balancing and managing data traffic while considering the vehicles using heterogeneous base stations. The approach can also deal with dynamic changes in the networks [31]. A novel approach named Hyperband uses the reinforcement learning algorithm which analyzes the speculative properties and also helps to speed up the random search through adaptive resource allocation. The proposed scheme has also been compared with the Bayesian optimization technique with respect to the issues related to the hyperparameter used to evaluate the performance of various machine learning algorithms.

13.2.3 UNSUPERVISED LEARNING ALGORITHM

A method based on the dynamic pattern for the location prediction is discussed in [32]. The study reveals the prediction of an individual's next location. The major challenge is to identify a more sophisticated model for the increase in the prediction of number of activity types to make the scheme versatile and more effective [33]. This study involves the machine learning unsupervised clustering algorithm to classify and identify the patterns for the reports of measurement succumbed to the base serving station using long-term evolution heterogeneous networks. The outcome of this method is much more accurate as compared to other methods with respect to the quality measure and automatic calculation of numeral of clusters with no preceding information. Conversely, the approach was not able to optimize the HO parameters for all the clusters individually recognized by the algorithm [34]. A machine learning approach was designed to advance the concert between internal femtocells and external microcells for long-term-evolution-based self-organizing networks. The scheme was built on the wireless setting and tries to overwhelm conferral in areas where it was executed unnecessarily, which further reduces unwanted signaling and results in increasing the overall effectiveness of the system [35]. A strong and efficient wireless localization algorithm has been given based on the relevance vector machine techniques to enhance the exactness in time-of-arrival localization to discover non-line-of-sight signals in the existing environment. This scheme presented an advanced version of Gaussian variational message passing to decrease the complexity in the computations while keeping the level of accuracy intact.

13.3   ISSUES RELATED TO THE PRESENT APPROACHES

Machine learning is considered as the most important artificial intelligence technology and thus has been widely used in the fields of multimedia security, image processing, pattern recognition, natural language processing, etc. The existing approaches of machine learning involve fewer requirements for implementation and computation, which has been considered as an important plus point. At the same time, however, it possesses some issues, mostly related to the authentication techniques based on the static system while coming across the composite and dynamic wireless setting. The summary of the challenges are discussed next.

13.3.1   INCONSISTENCY

The outcome of the authentication schemes based on a single attribute comes out to be less than perfect and varies with respect to the selected attribute. Additionally, the inadequate series of the allocation of the precise characteristic allocation may not be satisfactory for the distinguishing senders most of the time. This factor creates a hindrance in the performance of the authentication methods based on a single attribute in several worldwide areas of relevance, like maintaining the security of any mobile component.

13.3.2   OBSCURITY IN PRE-DESIGNING

The development of maximum authentication schemes at the physical layer involves the specific model to be the actual basis, and the same has been demonstrated in various articles [12,13]. On the other hand, when such model-based schemes are implemented in an intricate time-varying atmosphere, the result depreciates with a significant amount and the overall execution is affected. The working of the methods crucially requires a large amount of data with the complete awareness in the specific area to acquire particular model, which is perceptibly unwanted for the mobile networks like vehicular ad-hoc networks (VANETs). Thus, an important issue is to first design an accurate model of authentication for sustaining novel claims.

13.3.3 UNINTERRUPTED SECURITY TO GENUINE COMPONENTS

The authentication approaches already in use are usually static with respect to time, i.e., it does not support time-varying attribute and binary in environment, which specifies that either the component will be through or fail in the authentication process, finally directs to one-time verification with the increase in the difficulty level [12,13]. Several examples have been discussed in the literature where these schemes may not be able to detect a hitch even after following the authentication process of logging in as well as in other altering protection threats.

13.3.4 UNINTERRUPTED SECURITY TO GENUINE COMPONENTS: TIME-DIVERGENT FEATURES

With the movement of devices and variation in the time factor, the result may get affected as the performance of the authentication system alters randomly due to the lack of correlation between the attributes. Therefore, the divergence of the time-related characteristic amplifies the uncertainty of the challengers and, conversely, decreases the precision of authentication of genuine working physical components, irrespective of the knowledge of the different features of the system.

13.3.5 DEALING WITH VARIED NETWORK

As the latest advancements in technology, wireless networks have been efficiently handling the increase in the network traffic. In order to accommodate the issues related to this significant increase in intricacy, the user will have to recurrently toggle between the base centers and the access points which may often lead to compromise the verification process. Under these circumstances, it becomes more difficult to manage heterogeneous or varied networks. Conventionally, this handover was either due to key used during cryptography or on the basis of multiple handshakes, and at the same time, it engrosses numerous objects like web server, base center, access point, and end user, and this will lead to undesired issues.

13.3.6 INCORPORATING AUTHENTICATION PROTOCOLS

To implement the authentication techniques on the data sets, the major challenge is to amalgamate the present infrastructure with the set of rules and protocols. With this, there is also the problem of how to broaden the device connectivity for more specific situations of end-to-end authentication. But in huge wireless networks, the authentication process takes place between the different components which are not directly associated. On the other hand, the physical-layer authentication process is restricted to device-to-device verification as they rely on the straight relationship between the attributes of sender as well as receiver. Hence, it is crucial to extend an authentication process that is not limited to the physical layer of two directly related components. As per the previous discussion related to the current security issues and challenges, improvement in the efficient and useful authentication approaches are of great importance for all the advanced upcoming wired as well as wireless networks with the involvement of low-cost devices for areas of relevance in industry.

13.4 CONCLUSION

The chapter focused on the practical implication of authentication schemes based on artificial intelligence in wireless multimedia networks. The major consideration is to ensure the confidentiality of communication as well as to enable the authentication of various nodes containing multimedia types of documents. After the exhaustive literature review, it has found that the better security can be achieved at the physical layer. Finally, it is concluded that various issues still exist to securely transmit multimedia objects in a wireless environment. Therefore, there is constantly a scope for a deeper level of research work in this area.

REFERENCES

1. M. A. Matin and M. M. Islam. Overview of wireless sensor network, wireless sensor networks—technology and protocols. *IntechOpen* (September 6, 2012). doi:10.5772/49376.
2. S. P. Meenakshi, J. Thomas, K. S. Balaji, N. S. Narayanan, S. Raman, and V. Kamakoti. Detecting unauthorized access points in wireless environment. *International Journal of Computer Sciences and Engineering Open Access Survey Paper*, Vol. 6, Special Issue 11, December 2018, pp. 246–250, E-ISSN: 2347-2693.
3. S. Sandip, T. S. Vanjale, and P. B. Mane. A novel approach for fake access point detection and prevention in wireless network. *International Journal of Computer Science Engineering and Information Technology Research (IJCSEITR)*, Vol. 4, Issue 1, February 2014, pp. 35–42, ISSN (P): 2249-6831; ISSN (E): 2249-7943.
4. D. Kim, D. Shin, and D. Shin. *Unauthorized access point detection using machine learning algorithms for information protection*. In *2018 17th IEEE International Conference on Trust, Security and Privacy in Computing and Communications/12th IEEE International Conference on Big Data Science and Engineering*, New York, NY, USA, pp. 1876–1878, 2018.
5. M. Marks. A survey of multi-objective deployment in wireless sensor networks. *Journal of Telecommunications and Information Technology*, Vol. 3, 2010, pp. 36–41.
6. A. S. Papade, V. E. Pansare, R. D. Patil, and S. S. Gore. Unauthorized access point detection in wireless LAN. *International Journal of Advanced Research in Computer and Communication Engineering* Vol. 4, Issue 11, November 2015, pp. 371–373.
7. X. Qiu, Z. Du, and X. Sun, Artificial intelligence-based security authentication: Applications in wireless multimedia networks, special section on mobile multimedia: Methodology and applications, November 28, 2019, doi:10.1109/access.2019.2956480.
8. Z. Pan, C.-N. Yang, V. S. Sheng, N. Xiong, and W. Meng. Machine learning for wireless multimedia data security, *Hindawi Security and Communication Networks*, 2019.
9. Y. Sun, M. Peng, S. Y. Zhou, Y. Huang, and S. Mao. Application of machine learning in wireless networks: Key techniques and open issues. *IEEE Communications Surveys & Tutorials*, Vol. 21, Issue 4, March 2019, pp. 3072–3108.
10. H. Dai, H. Ren, Z. Chen, and G. Yang. Privacy-preserving sorting algorithms based on logistic map for clouds. *Security and Communication Networks*, Vol. 2018, Issue 1, September 2018, pp. 1–10.

11. B. Ma, B. Li, X.-Y. Wang, and C. Wang, Code division multiplexing and machine learning based reversible data hiding scheme for medical image, *Security and Communication Networks*, Vol. 2019, Issue 2, January 2019, pp. 1–9.

12. L. Fang, J. D. Wang, C. Ge, and R. Yongjun, Fuzzy conditional proxy re-encryption, *Science China: Information Sciences*, Vol. 56, Issue 5, May 2012, pp. 1–13.

13. Y. Zheng, J. Zhu, W. Fang, and L.-H. Chi, Deep learning hash for wireless multimedia image content security, *Hindawi Security and Communication Networks*, Vol. 2018, Article ID 8172725, 2018

14. Z. Zhang, Y. Li, C. Wang, M. Wang, Y. Tu, and J. Wang, An ensemble learning method for wireless multimedia device identification, *Machine Learning for Wireless Multimedia Data Security*, Vol. 2018, Article ID 5264526, 2018.

15. Y. Xue, W. Jia, X. Zhao, and W. Pang, An evolutionary computation based feature selection method for intrusion detection, *Machine Learning for Wireless Multimedia Data Security*, Vol. 2018, 2018.

16. S. Niu, R. Huang, W. Chen, and Y Xue, An improved permission management scheme of Android application based on machine learning, *Machine Learning for Wireless Multimedia Data Security*, Vol. 2018, 2018.

17. P. Zhang and W Zhang, Differential cryptanalysis on block cipher skinny with MILP program, *Machine Learning for Wireless Multimedia Data Security*, Vol. 2018, 2018.

18. R. Yi, M. Nie, S. Li, and C. Li, "Single image de-raining via improved generative adversarial nets, *Sensors(Basel)* Vol. 20, Issue 6, March 2020, p. 1591. doi:10.3390/s20061591.

19. X. Zhao, H. Mao, S. Liu, and W. Song, Analysis on matrix GSW-FHE and optimizing bootstrapping, *Security and Communication Networks* Vol. 2018, Issue 1, December 2018, pp. 1–9.

20. Z. Zhang, X. Zhou, X. Zhang, L. Wang, and P. Wang, A model based on convolutional neural network for online transaction fraud detection, Vol. 2018, 2018.

21. E. Min, J. Long, Q. Liu, J. Cui, and W. Chen, TR-IDS: Anomaly-based intrusion detection through text-convolutional neural network and random forest, *Machine Learning for Wireless Multimedia Data Security*, Vol. 2018, 2018, Article ID 4943509.

22. Y. E. Sagduyu, Y. Shi, T. Erpek, W. Headley, B. Flowers, G. Stantchev, and Z. Lu, When wireless security meets machine learning: Motivation, challenges, and research directions. January 2020.

23. R. Boutaba, M. A. Salahuddin, N. Limam, S. Ayoubi, N. Shahriar, F. Estrada-Solano, and O. M. Caicedo, A comprehensive survey on machine learning for networking: Evolution, applications and research opportunities, *Journal of Internet Services and Applications* Vol. 9, 2018, p. 16. doi:10.1186/s13174-018-0087-2.

24. B. Choi, J. Kim, and J. Ryou, *Retrieval of illegal and objectionable multimedia, Fourth International Conference on Networked Computing and Advanced Information Management*, IEEE, 2008.

25. H. Fang, X. Wang, and S. Tomasin, Machine learning for intelligent authentication in 5G and beyond wireless networks, *IEEE Wireless Communications*, October 2019.

26. L. Xu and A. Nallanathan, Energy-efficient chance-constrained resource allocation for multicast cognitive OFDM network, *IEEE Journal on Selected Areas in Communications*, 2016. doi:10.1109/JSAC.2016.2520180.

27. M. Lin, J. Ouyang, and W.-P. Zhu, Joint beamforming and power control for device-to-device communications underlaying cellular networks, *IEEE Journal on Selected Areas in Communications, IEEE Journal on Selected Areas In Communications*, September 2015. doi:10.1109/JSAC.2015.2452491.

28. S. Samarakoon, M. Bennis, W. Saad, and M. Latva-Aho, Dynamic clustering and ON/OFF strategies for wireless small cell networks, arXiv:1511.08631v1 cs.NI. November 27, 2015.

29. Q. Zhao, D. Grace, A. Vilhar, and T. Javornik, *Using K-means clustering with transfer and Q learning for spectrum, load and energy optimization in opportunistic mobile broadband networks*, *2015 International Symposium on Wireless Communication Systems (ISWCS)*, pp. 116–120, 2015. doi:10.1109/ISWCS.2015.7454310

30. Z. Li, C. Wang, and C.-J. Jiang, User association for load balancing in vehicular networks: An online reinforcement learning approach, *IEEE Transactions on Intelligent Transportation Systems*, 2017.

31. L. Li, K. Jamieson, G. DeSalvo, A. Rostamizadeh, and A. Talwalkar, Hyperband: A novel bandit-based approach to hyperparameter optimization, *Journal of Machine Learning Research* Vol. 18, pp. 1–52, 2018.

32. C. Yu, Y. Liu, D. Yao, L.T. Yang, H. Jin, H. Chen, and Q. Ding, Modeling user activity patterns for next-place prediction, *IEEE Systems Journal*, Vol. 11, Issue 2, pp. 1060–1071, 2015.

33. D. Castro-Hernandez and R. Paranjape. Classification of user trajectories in LTE HetNets using unsupervised shapelets and multi-resolution wavelet decomposition. *IEEE Transactions on Vehicular Technology*, Vol. 66, Issue 9, pp. 7934–7946, 2017.

34. N. Sinclair, D. Harle, I. A. Glover, J. Irvine, and R. C. Atkinson. An advanced SOM algorithm applied to handover management within LTE. *IEEE Transactions on Vehicular Technology*, Vol. 62, Issue 5, June 2013.

35. T. V . Nguyen, Y. Jeong, H. Shin, and M. Z. Win. Machine learning for wideband localization. *IEEE Journal on Selected Areas in Communications*, Vol. 33, Issue 7, pp. 1357–1380, 2015.

14 Low Power Communication in Wireless Sensor Networks and IoT

Pawan Kumar Sharma, Jaspreet Singh, Yogita, and Vipin Pal

National Institute of Technology, Meghalaya, India

CONTENTS

14.1 INTRODUCTION

Data communication is the phenomenon by which two or more devices can interact with each other to share information. In communication, data can be exchanged through transmission medium—wired (guided) and wireless (unguided). In wired communication, the data is transmitted over a wired communication technology like coaxial cable or optical fiber. The wired communication started first in 1874 in which the two communicating devices were connected through twisted-pair cables that transmit the audio signals from the caller to the receiver. The conduit in wired transmission includes twisted-pair cable, coaxial cable, and fiber-optic cable. The transmission rate of the media is limited by the physical characteristics of the conduit. On the other hand, in wireless data communication, data in the form of electromagnetic waves is transmitted through free space without using a physical conduit. The electromagnetic spread spectrum, as shown in Figure 14.1, is a wide array of frequencies including gamma rays, X rays, UV, visible, IR and radio whose frequency lies between 2.4 and 2.4835 GHz, with

DOI: 10.1201/9781003145028-14

the advantage of interference-free communication being coordinated and systematized for specific geographical areas. Licensed spectrum wireless communication is highly dependent on infrastructures like cellular networks, radio networks, and others. On the other hand, an unlicensed spectrum, with frequencies at 902–928 megahertz (MHz), 2400–2483.5 MHz, and 5725–5850 MHz, is free to use and meant for short-range communication like a personal area network (PAN). The wireless communication setup for unlicensed spectrum is quick, easy to maintain, and cost-effective. Consequently, the unlicensed spectrum is well suited for the infrastructure-less networks.

Wireless sensor networks and Internet of Things are infrastructure-less networks and operate on an unlicensed spectrum. They have been the captivating interest of researchers because of their wide range of applications. Wireless sensor networks have been set up to monitor crop health at low cost for precision agriculture [1,2]. In [3], wireless sensor networks have been deployed to detect the oil spillage from onshore and offshore locations. The deployed WSNs ensure the safety and security in the oil and gas companies to further enhance environmental safety and economic growth. Home automation is one of the most widely deployed applications of IoT [4]. IoT has a vast area of applications such as healthcare—monitoring a patient in real time [5], smart city-monitor traffic conditions [6], water distribution [7], waste management [8], smart homes–temperature control, security, and many more [9].

Wireless sensor networks (WSNs) are groups of sensor nodes working in collaboration to monitor physical phenomenon from the region of interest [10]. WSNs involve a great number of sensor nodes, which are small in size with limited resources such as battery, computational power, communication range, and at least one base station which can be located inside or outside the region of interest. The sensor nodes collect the data about the physical situation and send it to the base station directly or via other sensor nodes as relay for further processing [11,12]. A WSN deployment architecture is shown in Figure 14.2.

Figure 14.3 demonstrates the various parts like sensors, microcontroller, transceivers, and memory and power source of the sensor nodes. Regarding power source, the power can be provided by the battery to other parts of the nodes to complete the processing. Sensing any physical phenomenon consumes energy, and handling data within the node consumes energy in microcontrollers. The radio transceiver, on the other

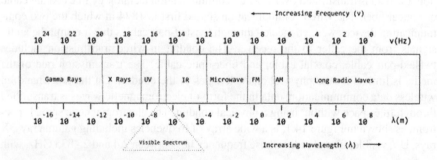

FIGURE 14.1   Electromagnetic spread spectrum.

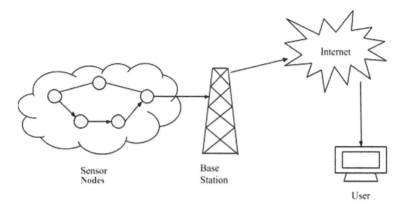

FIGURE 14.2 Wireless sensor network architecture.

hand, consumes maximum amount of energy while transmitting and receiving data during communication. Last but not least, computation of the data also consumes energy. The IoT usually deals with enormous data in home automation, industry, transportation, building automation, logistics, etc. A lot of energy is consumed handling this humongous data. The battery starts depleting during communication and gets exhausted. In addition, removing or recharging the battery is not entirely possible in most applications, so it is necessary to use the battery power efficiently and cost-effectively to maximize battery life. Low-power communication can still be achieved through following operations. Radio optimization saves energy by optimizing in transmission, modulation, cooperative communication, and directional antennas [13]. Data reduction uses aggregation of data, adapting sampling, compression, and network coding to save

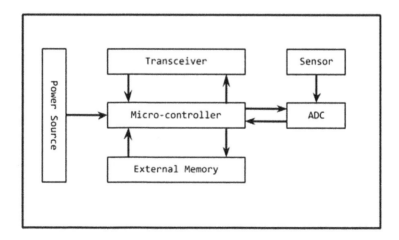

FIGURE 14.3 Sensor node architecture.

energy [14]. Sleep/waking schemes [15] and routing [16] are both used to save energy. Battery change and wireless energy transfer are also used to save power [17].

The Low Power Wide Area Network (LPWAN) like LoRa, Zigbee, SIGFOX, 6LoWPAN, NBIoT, etc., is used to save power. LPWAN consumes low power in communication which is desired in WSN and IoT to save energy. The work of this chapter describes the various new technologies for the low-power communication in WSN and IoT with their application. Section 14.2 describes long range based communication (LoRa). Section 14.3 describes about the Zigbee-based communication. In Section 14.4, 6LoWPAN is described. Sections 14.5 and 14.6 describes NBIoT- and SIGFOX-based communication. In Section 14.7, the conclusion of the work is described.

14.2 LONG RANGE COMMUNICATION (LORA)

LoRa is one of the LPWAN protocols used to save energy in WSN and IoT. It is a modulation technique built in physical layer to modulate signals. LoRa uses Chirp Spread Spectrum (CSS) modulation. A chirp in CSS refers to a signal with a continuously increased or decreased frequency that sweeps and wraps around a predefined bandwidth, called up-chirps and down-chirps [18]. LoRa is based on the modulation of the Chirp Spread Spectrum (CSS). The CSS, first proposed by Winkler [19], is a subcategory of Direct-Sequence Spread Spectrum. In this spreading technique, a symbol is encoded with a large sequence of bits which eventually reduces the signal to noise ratio at the receiver end without changing the bandwidth of wireless signal. The length of this spreading code is equal to 2SF, where SF is adjustable parameter known as spreading factor that can vary from 7 to 12 by which the data rate, throughput, coverage range and energy consumption vary. It adopts a duty cycle transmission that can exchange the data for a longer time and produce a large distance communication [18,20–22].

LoRa has revolutionized IoT by allowing long-range data communication with much less power. LoRa is versatile in smart cities, smart homes and buildings, smart infrastructure, intelligent metering, and smart supply chain and logistics for rural and indoor applications. The PHY layer of LoRa is proprietary; the rest of the protocol is known as LoRaWAN. The three main building blocks of the LoRaWAN network are LoRa end-nodes, gateways, and a network server. The end nodes are used to collect the information of any physical phenomenon. The collected data from the end nodes are transferred to the gateways. Only one gateway is used to collect the data form the end nodes to save the power in communication. These data are now sent to the network server for further processing [18].

H. Huh et al. [23] analyzed the standard LoRa-based networks in IoT applications. The main purpose of the proposed work is to overcome the shortcomings of the standard LoRaWAN. In the actual LoRaWAN topology, direct communication between the nodes is not possible. Due to this reason, efficient routing of the packets to the gateways could not be achieved. As a result, mesh topology is adopted in the proposed work in which each node transfers a packet by communicating to different nodes in order to effectively route the data to the base station. Because of this mutual cooperation and node interoperability, optimum coverage has been achieved, which is a difficult task in a network based on a star topology.

M. Babazadeh et al. in [24] used LoRa in a smart water network for edge-based anomaly detection. In this work, an architecture with 10 sensor nodes was proposed, where a base station or a central processor is used to collect the data from various sensors. These sensors have the information of the leakage in the pipelines. Here, the local sensor nodes sense the nearby water leakage through water pressure variations which may be caused by sudden burst or leakage in the pipelines. A Arduino 101 low power board is used for all the SNs, which consists of LoRa shield ($S X 1276 M B 1 M A S$) that provides the wide area communication and reduces the power consumption in the network.

J. Wang et al. [25] implemented a hardware setup for small monitoring the environment, in which WIFI LoRa 32 is used as end node and gateways. A WSN network is formed to monitor the environment. The end nodes were used to detect the environment condition followed by the gateways and then MQQT server. The fire is detected using this small setup.

V. K. Sarker et al. [26] investigated various applications of IoT with LoRa, such as smart cities, smart metering, etc. Various sensors were deployed to collect the information of environment at the city administrators. The chemical industries sensors are deployed to monitor temperature, humidity, and electrochemical gas. This architecture consists of an edge of extra layer. The edge-assisted gateway is used to compress the data using lossy and lossless algorithms depending upon the application. The collected data are then passing to a LoRa gateway. In this way, an efficient utilization of bandwidth has been achieved. Further, this information is processed by edge-assisted gateways to extract the relevant data which can be sent to LoRa gateways for saving the power resource.

Environmental monitoring is a major research area that requires an autonomous and robust physical environment data collection system without any physical interaction. A heterogeneous multiprocessor sensor platform of an ultralow-power microcontroller, the Strong performance processor, is used to monitor the environment [27]. Multimedia data such as audio and video are collected by the nodes that require a huge amount of energy. A node consists of a low-power microprocessor (LPM), multimedia processor (MP), multimedia sensors, radio module (RM), traditional sensors, and power supply. To monitor such a large area, LoRa in star topology is used for low-power, long-range and low-data rate communication. LPM is used in sensors to detect the data from the environment to store it temporarily. With the help of LPM, MP is used to detect the information from the environment, and using the radio module, it informs the other nodes and a base station for relevant information.

As power is the main concern in all the applications like monitoring the industries, environment, home automation etc., LoRa is effectively used to provide low-power and long-distance communication in the WSN and IoT.

14.3   ZIGBEE

Zigbee is used as LPWAN, a standard communications protocol built on top of the 802.15.4 protocol added mainly for low-power communication and wireless mesh networking. The devices which support the Zigbee protocol operate mostly in power-saving mode. Due to this reason, Zigbee has been considered as a standard for low-power communication [28]. Like any communication protocol, Zigbee employs the concept of layers to separate different components into independent modules so they can operate without any intervention. The physical layer and MAC layer have been defined by IEEE 802.15.4, and the topmost layers, including the application layer and network layer, are defined by Zigbee. It is clear from this fact that any Zigbee-compliant device must conform to IEEE 802.15.4 as well [29].

Zigbee has three devices, Zigbee Controller (ZC), Zigbee Router (ZR), and Zigbee End Device (ZED) [30]. Zigbee Controller is the network's most competent device responsible for network setup and manages network information and security keys. On the other hand, Zigbee Router is responsible for transmitting data from device to device, and Zigbee End Device has sufficient technique to talk to the parent device. [31]. Zigbee currently supports two types of networks—the non-beacon-enabled network and the beacon-enabled network. Non-beacon-enabled networks are types of heterogeneous networks in which some devices consume power continuously, while others transmit data only when some external trigger is applied to them. Unlike non-beacon-enabled networks, routers relay periodic beacons to monitor their presence in the beacon-enabled networks. It ensures that the routers do not need to be constantly involved, and the remaining nodes can sleep in between the beacons and prolong their battery life [32].

Khusvinder Gill et al. [32] proposed a Zigbee-based low-cost architecture in home automation. Figure 14.4 shows the standard Zigbee-based home automation architecture in which the Zigbee Coordinator is responsible for managing all the Zigbee devices within the network. The communication between the two end devices takes place through the coordinator. Zigbee's wireless nature helps conquer the installation of Zigbee-compliant devices within the existing infrastructure. Not only that, Zigbee theoretically provides a 250 kbps data rate which is enough to control most of the home automation devices as the requirement is not greater than 40 Kbps.

Basically, four steps are provided in the presented work. In the first step, remote users will start controlling the home devices through the internet. In the second step, commands from the remote users are passed to the home gateway through the internet. Then, to ensure the safety and security of the system, all commands will pass to the virtual home and it will then eventually reach home devices.

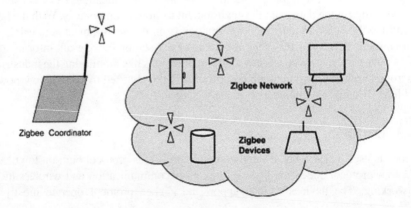

FIGURE 14.4 Zigbee-based home automation architecture.

Zhang Qian et al. [29] provides the solution for greenhouse monitoring. The proposed Zigbee network consists of sensor nodes such as temperature sensors, humidity sensors, and light sensors. The Zigbee module is used to gather environmental data and relay it to the handheld controller (HHC) which consists of ARM MCU and the Zigbee module. Now, the data are stored in HHC and shown on the monitor. Afterward, the HHC sends the control to the actuator. All the used sensor nodes are based on the Zigbee module.

M. Keshtgari et al. [33] provide the precision agriculture to improve crop production based on Zigbee technology. This work focuses mainly on the two topologies of network. In the first, each sensor is deployed at the corner of the grid, and the server node is placed in the middle. The server node or gateway receives the data from the nearby Zigbee sensor nodes. A different IP address is used to connect the sensor nodes to the base station. Thus, the whole area is managed and controlled. In the second scenario, the server node or base station is placed out of the area. The average range of 10 m stems from sensor ranges where there is a high contact probability between them that consists of the signal-to-noise ratio higher with threshold. Crop production is monitored by these two topologies.

In general, the Zigbee protocol minimizes the time in which the radio is on, thus maximizing battery life. In the beacon-enabled network, the nodes need to be activated when the beacon is transmitted, while in the non-beacon-enabled network, some devices are always active while others will remain in sleep mode for a long time.

14.4 IPV6 LOW POWER PERSONAL AREA NETWORK (6LOWPAN)

An underdeveloped standard from the Internet Engineering Task Force (IETF), 6LoWPAN is based on the IPv6 that allows data transfer in a specific mesh or star network. It is widely used in the WSN and IoT [34]. Based on the IEEE 802.15.4 communication protocol, 6LoWPAN found its application in home and building automation, healthcare automation, industrial automation, real-time environment monitoring, vehicular automation, etc. The reason why 6LoWPAN is popular is that it is based on an IP-based device that can easily be connected to any IP network which is open and free to use [35,36].

In 6LoWPAN, low-power and lossy networks (LLNs) development is generally focused on energy saving, which in most cases is a very restricted asset. One way to save power is to control the amount of data transmitted. The frame size is limited to much smaller values in a range of LLN standards than the promised 1280-byte IPv6 Maximum Transmission Unit (MTU). In general, an LLN based on IEEE 802.15.4 classical physical layer (PHY) is limited to 127 bytes per frame. The need to compact IPv6 packets over IEEE 802.15.4 has contributed to the writing of IEEE 802.15.4-based networks compression format for IPv6 datagrams. Innovative route-over strategies for routing within an LLN have been and are still being developed. Therefore, the network addresses limited packet size and low bandwidth, and it needs energy savings to sustain network node life [37].

M. S. Shahamabadi et al. [37] proposed a 6LoWPAN for monitoring the patients in the hospital with the help of WSN and IoT. Mobility support in 6LoWPAN is still in its infancy, and it is important to maintain proper mobility to monitor the patient locations in the hospital network. In the proposed work, the authors survey the IPv6 mobility protocols and proposed solution to better fit in the hospital architecture. The structure of the hospital system consists of patients with mobile nodes and sensors, the Monere system which consists of local gateway or border router, Internet gateway, Hospital Information System (HIS) and users like physicians, surgeons, and nurses. All the nodes in PAN are deployed with 6LowPAN. All the information about the patient is obtained by the sensor nodes. This data can be handed over to the local gateway through the border router with the help of 6LoWPAN, so that it consumes less power. Further, this data can be sent to the doctor via main gateway, so that the doctor can take the necessary action.

Dhananjay Singh et al. [38] proposed a real-time monitoring of patients that is designed for global healthcare monitoring. The biomedical sensor with 6LoWPAN nodes is fixed for the body area network. In the hospital or private area networks, patients can move freely. That 6LoWPAN node has its own IP address for the use of multi-hop, mesh routing between nodes and independently wireless internet or gateway connections. The service provider pings or connects directly with the patient using internet provider equipment and collects the patient's current status. The received data from the body sensors are collected and sent to the doctor via the internet using 6LoWPAN that consumes low power in processing.

Zucheng Huang et al. [39] discussed the smart lighting application using 6LoWPAN. The architecture of the smart lighting system comprises three layers: 6LoWPAN sensing layer, processing layer, and application layer. The sensing layer is fabricated by LED lights, light controllers, and sensors. LED is one of the most energy-saving and adjustable dimming street lights. The light controller functions in the network as terminal nodes and then reports data and/or events via a PLC connected to the data server. Sensors are used to perceive light sense and shift objects. The information process layer involves a centralized controller, database, and related connection to the transmission. For the collection of data from a street block, hierarchical controllers are mounted in switch gears. The 6LoWPAN light controllers replaced the PLC light controllers, and border routers replaced the centralized controllers. For the collection of data from all terminal nodes within their contact range, the main controllers are deployed in each street block. The hierarchical controllers relay PLC packets via the internet to the data server.

Mainly, the incorporation of LLN and smaller frame size leads to low power consumption in 6LoWPAN.

14.5 NARROW BAND INTERNET OF THINGS (NBIOT)

NBIoT uses a cellular network to connect an enormous number of smart devices like cameras, wearable devices, etc. [40]. NBIoT is suitable for stationary, low-power and long-range applications such as smart metering, smart agriculture, and smart city applications. NBIoT is a standard narrowband system that does not use a conventional physical LTE layer but a subset of it. It limits the bandwidth to a 200 KHz narrow band (which is why it called NBIoT) and therefore gives it a longer range and low performance compared to LTE-M and LTE. The downlink is 60 Kbps and uplink is 30 Kbps. It is suitable for applications needing low-power static applications [41].

The NBIoT reduces power consumption and increases battery lifetime. Two methods are involved in reducing power consumption—discontinuous reception (DRX) and power-saving mode (PSM). In DRX, the user equipment discontinuously receives the physical downlink control channel (PDCCH). It configures through DRX cycles. Each DRX cycle has two phases. First, the user monitors the PDCCH for a short period, Second, it stops monitoring PDCCH for a long period. It manages the sleep, wakeup, and communication very effectively. The advantage of DRX is that it helps the 'N' number (40+) of hyper frames to occur at 10.24 seconds each before a device wakes up in its next paging window for activity again. DRX allows computers to sleep and communicate in a highly efficient and coordinated way, thus saving the power of the node. In PSM, it allows the node to switch to deep sleep mode when it is idle, and it remains in the deep sleep mode until some mobile-originated transactions wake it up. Thus, NBIoT promotes longer battery life [42]. It has a vast area of applications including smart healthcare, smart home automation, smart metering, intelligent transportation systems, and many more [43].

Xiaojun Wu et al. [43] proposed the architecture of WSN in which NBIoT consumes low energy with a well-implemented security mechanism to avoid malware infection. In traditional WSNs, security is the main issue. A large number of nodes in the traditional WSN are homogeneous with antimalware abilities. Thus, the guarantee of the node destroyed by the virus depends upon the node information exchange behavior. Hybrid network nodes are heterogeneous. The attacked malicious nodes and available nodes are determined on the basis of their degree. WSN nodes' availability is defined as the node's probability to operate normally. Evaluating the node availability is one of the key issues of the NBIoT-HWSN (NBIoT Heterogeneous WSN). In the presented work, the transitions between the states of the heterogeneous nodes are represented using the epidemiological theory and the Markov chain. The node availability equation is obtained by the estimate.

Jiong Shi et al. [44] used NBIoT in smart parking. The proposed system is composed of four subsystems. First, we have a parking lot with sensor nodes which is formed from the BC95 NBIoT module, STM32F103 MCU and geomagnetic vehicle detector. BC95 provides ultra-low-power consumption and data passing services to the end user; thus it is considered the best choice for a wide area of IoT applications. Second, we have a smart parking cloud server in which the collected data from the sensor node are stored. Third, we have an application for mobile devices which is divided into two subcategories: platform-layer substructure and web-app layer substructure. Web-app-layer substructure is built based on a platform-layer substructure. Lastly, we have a third-party payment platform.

NBIoT limits the bandwidth to a narrow band of just 200 KHz with low uplink and downlink. Thus, it is considered a good choice for low-power applications.

14.6 SIGFOX

SIGFOX is a IoT network that is used to transmit data without the need for network connections. In this case, there is no overhead during communication. SIGFOX provides a software-based networking system that handles all the functionality of the network and computation in the cloud. Altogether, it dramatically reduces connected devices' energy consumption and costs.

SIGFOX provides a lightweight protocol that facilitates the transmission of small-size messages. Modern protocols fully support the transmission of large amounts of data but are highly ineffective in the transmission of small sparse data. SIGFOX with its new lightweight protocol has a frame size of 26 bytes for a 12-byte data in it. For comparison, TCP/IP stack has a frame size of 40 bytes for a 12-byte data in it. Lighter protocol frame size therefore results in less data to be sent and ultimately leads to less energy consumption [45]. SIGFOX is an LPWAN used for various applications in WSN and IoT. SIGFOX is used to broadcast the data without maintaining the network connection [46]. It offers an end-to-end connectivity solution for IoT. Base stations have software-defined radios connected through the IP-based network. It has low noise, low power consumption, huge sensitivity to the receiver, and a cheap-cost antenna design. Initially it was capable only of uplinked transmission, but later, bidirectional communication was developed. The downlink transmission takes place only after the transmission of an uplink.

D. M. Hernandez et al. [46] proposed an LPWAN for the Industrial IoT. SIGFOX pokes out as an ultra-narrow band that enables very little of their power requirements. SIGFOX provides an end-to-end LPWAN solution, either by itself or in collaboration with other network operators. The proposed topology consists of three interconnected networks, namely Star Topology. In addition to SIGFOX mode, which provides access to the SIGFOX network, two different networking modes are used by the nodes: P2P mode, in which nodes are used to exchange information directly without using the SIGFOX network; and hybrid mode, a combination of P2P and SIGFOX that is used for communication.

Thomas Janssen et al. [47] proposed outdoor fingerprinting localization using SIGFOX. Fingerprinting is a localization technique that utilizes the Received Signal Strength Indicator (RSSI) to find the position of the sending device. A fingerprinting database is built in the first offline process, which holds RSSI measurements of training sites in a known region. A wireless transmitter can be identified during the online process by comparing the RSSI measurements in real time. RSSI measurements are used in their implementation to calculate the distance between a SIGFOX transmitter and the answering gateways for the transmitter's location in a considered area. In such a category, the position estimate is enhanced by calculating the range between end devices and GPS. SIGFOX uses the lightweight protocol to reduce the payload size; hence, low communication power is consumed.

14.7   CONCLUSION

This chapter described the various LPWAN technique including LoRa, Zigbee, 6LoWPAN, NBIoT, and SIGFOX with their applications in achieving low-power communication. As in LoRa, physical layer modulation CSS reduces power consumption. In Zigbee, nodes must stay in sleep mode until some beacon in the beacon-enabled network is activated or some external signal in the non-beacon-enabled network is applied. In 6LoWPAN and SIGFOX, reduced frame size can increase the node lifetime, while in NBIoT, narrow band technology saves power as it is focused on limiting the bandwidth to a single narrow band.

ACKNOWLEDGMENT

The work of the chapter has been supported by the National Mission on Himalayan Studies (NMHS) by the Ministry of Environment, Forest & Climate Change (MoEF&CC) sanctioned project titled "Cloud-assisted Data Analytics based Real-Time Monitoring and Detection of Water Leakage in Transmission Pipelines using Wireless Sensor Network for Hilly Region" (Ref. No. GBPNI/NMHS-2017-18/SG21).

REFERENCES

1. A. Rehman, A. Abbasi, N. Islam, and Z. Shaikh, "A review of wireless sensors and networks applications in agriculture," *Computer Standards and Interfaces*, vol. 36, 02, pp. 263–270, 2014.
2. S. M. T. Ojha and N. S. Raghuwanshi, "Wireless sensor networks for agriculture: The state-of-the-art in practice and future challenges," *Computers and Electronics in Agriculture*, vol. 118, pp. 66–84, 2015.
3. B. M. Sahoo, R. K. Rout, S. Umer, and H. M. Pandey. *"ANT Colony Optimization based optimal path Selection and data gathering in WSN,"* in *2020 International Conference on Computation, Automation and Knowledge Management (ICCAKM)*, pp. 113–119. IEEE, 2020.
4. M. Z. Abbas, K. A. Bakar, M. Ayaz, M. H. Mohammed, and M. Tariq, "Hop-by-hop dynamic addressing based routing protocol for monitoring of long range underwater pipeline," *TIIS*, vol. 11, pp. 731–763, 2017.
5. D. Evans, "The internet of things how the next evolution of the internet is changing everything," *CISCO White Paper*, vol. 1, pp. 1–11, 2011.
6. N. V. R. Kumar, *"IoT architecture and system design for healthcare systems,"* in *2017 International Conference on Smart Technologies for Smart Nation (SmartTechCon)*, pp. 1118–1123, 2017.
7. N. B. Soni and J. Saraswat, *"A review of iot devices for traffic management system,"* in *2017 International Conference on Intelligent Sustainable Systems (ICISS)*, pp. 1052–1055, December 2017.
8. M. Suresh, U. Muthukumar, and J. Chandapillai, *"A novel smart water-meter based on iot and smartphone app for city distribution management,"* in *2017 IEEE Region 10 Symposium (TENSYMP)*, pp. 1–5, July 2017.
9. A. Khan and A. Khachane, *"Survey on iot in waste management system,"* in *2018 2nd International Conference on I-SMAC (IoT in Social, Mobile, Analytics and Cloud) (I-SMAC)I-SMAC (IoT in Social, Mobile, Analytics and Cloud) (I-SMAC), 2018 2nd International Conference on*, pp. 27–29, Aug 2018.
10. A. C. Jose and R. Malekian, "Improving smart home security: Integrating logical sensing into smart home," *IEEE Sensors Journal*, vol. 17, pp. 4269–4286, July 2017.
11. S. R. Ibrahiem M. M. El Emary, *Wireless Sensor Networks: From Theory to Applications*. CRC Press, 1st ed., 2013.
12. F. Hu, *Wireless Sensor Networks: Principles and Practice*. CRC Press, 1st ed., 2010.
13. S. Vishwakarma, P. Upadhyaya, B. Kumari, and A. Mishra, "Smart energy efficient home automation system using IoT," in *2019 4th International Conference on Internet of Things: Smart Innovation and Usages (IoT-SIU)*, pp. 1–4, 2019.

14. E. Kranakis, D. Krizanc, and E. Williams, *"Directional versus omnidirectional antennas for energy consumption and k-connectivity of networks of sensors,"* in *OPODIS*, 2004.
15. I. Hou, Y. Tsai, T. F. Abdelzaher, and I. Gupta, *"Adapcode: Adaptive network coding for code updates in wireless sensor networks,"* in *IEEE INFOCOM 2008—the 27th Conference on Computer Communications*, pp. 1517–1525, April 2008.
16. J. Hsu, S. Zahedi, A. Kansal, M. Srivastava, and V. Raghunathan, *"Adaptive duty cycling for energy harvesting systems,"* in *ISLPED '06 Proceedings of the 2006 International Symposium on Low Power Electronics and Design*, pp. 180–185, October 2006.
17. Z. Wang, E. Bulut, and B. K. Szymanski, *"Energy efficient collision aware multipath routing for wireless sensor networks,"* in *2009 IEEE International Conference on Communications*, pp. 1–5, June 2009.
18. C. A. Trasviña-Moreno, R. Blasco, R. Casas, and A. Asensio, *"A network performance analysis of LoRa modulation for LPWAN sensor devices,"* In *Ubiquitous Computing and Ambient Intelligence*, pp. 174–181. Springer, Cham, 2016.
19. M. Winkler, "Chirp signal for communication,"*IEEE WESCON Convention Record* p. 7, 1962.
20. I. Butun, N. Pereira, and M. Gidlund, "Security risk analysis of LoRaWAN and future directions," *Future Internet*, vol. 11, p. 3, 2018.
21. M. H. Eldefrawy, I. Butun, N. Pereira, and M. Gidlund, "Formal security analysis of Lo-RaWAN," *Computer Networks*, vol. 148, pp. 328–339, 2019.
22. F. M. Costa and H. Ochiai, *"Energy-efficient physical layer design for wireless sensor network links,"* in *2011 IEEE International Conference on Communications (ICC)*, pp. 1–5, June 2011.
23. H. Huh and J. Kim, *"LoRa-based mesh network for IoT applications,"* in *2019 IEEE 5th World Forum on Internet of Things (WF-IoT)*, pp. 524–527, 04 2019.
24. M. Babazadeh, "Edge analytics for anomaly detection in water networks by anarduino101-LoRa based WSN," *ISA Transactions*, vol. 92, pp. 273–285, 02 2019.
25. J. Wang, S. Yi, D. Zhan, and W. Zhang, *"Design and implementation of small monitoring wireless network system based on LoRa,"* in *2019 IEEE 4th Advanced Information Technology, Electronic and Automation Control Conference (IAEAC)*, vol. 1, pp. 296–299, December 2019.
26. V. K. Sarker, J. P. Queralta, T. N. Gia, H. Tenhunen, and T. Westerlund, *"A survey on lora for IoT: Integrating edge computing,"* in *2019 Fourth International Conference on Fog and Mobile Edge Computing (FMEC)*, pp. 295–300, June 2019.
27. S. Garca, D. F. Larios, J. Barbancho, E. Personal, J. M. Mora-Merchan, and C. León, "Heterogeneous lora-based wireless multimedia sensor network multiprocessor platform for environmental monitoring," *Sensors*, vol. 19, no. 16, p. 3446, 2019.
28. P. Kinney, *ZigBee Technology Wireless Control that Simply Works*. 1st ed., 2003.
29. Q. Zhang, X.-l. Yang, Y.-m. Zhou, L.-r. Wang, and X.-s. Guo, "A wireless solution for greenhouse monitoring and control system based on ZigBee technology," *Journal of Zhejiang University-Science A*, vol. 8, pp. 1584–1587, October 2007.
30. C. M. Ramya, M. Shanmugaraj, and R. Prabakaran, *"Study on ZigBee technology,"* in *2011 3rd International Conference on Electronics Computer Technology*, vol. 6, pp. 297–301, April 2011.
31. W. Wang, G. He, and J. Wan, *"Research on ZigBee wireless communication Technology,"* in *2011 International Conference on Electrical and Control Engineering*, pp. 1245–1249, September 2011.
32. K. Gill, S.-H. Yang, F. Yao, and X. Lu, "A ZigBee-based home automation system," *Consumer Electronics, IEEE Transactions on*, vol. 55, pp. 422–430, 06 2009.

33. M. Keshtgari and A. Deljoo, "A wireless sensor network solution for precision agriculture based on zigbee technology," *Wireless Sensor Network*, vol. 4 01, pp. 25–30, 2012.
34. K. McGee and M. Collier, "*6LoWPAN forwarding techniques for IoT*," in *2019 IEEE 5th World Forum on Internet of Things (WF-IoT)*, pp. 888–893, April 2019.
35. R. Herrero, "6LoWPAN fragmentation in the context of IoT based media rtc," 2019.
36. L. Anhtuan, J. Loo, A. Lasebae, M. Aiash, and Y. Luo, "6LoWPAN: A study on QoS security threats and countermeasures using intrusion detection system approach,"*The International Journal of Communication Systems*, vol. 25, pp. 1189–1212, Sept. 2012.
37. M. S. Shahamabadi, B. B. M. Ali, P. Varahram, and A. J. Jara, "*A network mobility solution based on 6LoWPAN hospital wireless sensor network (nemo-hwsn),*" in *2013 Seventh International Conference on Innovative Mobile and Internet Services in Ubiquitous Computing*, pp. 433–438, July 2013.
38. D. Singh, U. S. Tiwary, H.-J. Lee, and W.-Y. Chung, "*Global healthcare monitoring system using 6LoWPAN networks,*" in *Proceedings of the 11th International Conference on Advanced Communication Technology - Volume 1, ICACT'09*, (Piscataway, NJ, USA), pp. 113–117, IEEE Press, 2009.
39. Z. Huang and F. Yuan, "Implementation of 6LoWPAN and its application in smart lighting," *Journal of Computer and Communications*, vol. 03, pp. 80–85, 01 2015.
40. P. Andres-Maldonado, P. Ameigeiras, J. Prados-Garzon, J. Navarro-Ortiz, and J. M. López-Soler, "Narrowband IoT data transmission procedures for massive machine-type communications," *IEEE Network*, vol. 31, pp. 8–15, 2017.
41. A. Sultania, C. Delgado, and J. Famaey, "Implementation of NB-IoT power saving schemes in ns-3," June 2019.
42. P. Sakshi and S. Jain, "A survey on energy efficient narrowband internet of things (NBIoT): Architecture, application and challenges," *IEEE Access*, vol. 7, pp. 16739–16776, 2018.
43. X. Wu, Q. Cao, J. Jin, Y. Li, and H. Zhang, "Nodes availability analysis of NB-IoT based heterogeneous wireless sensor networks under malware infection,"*Wireless Communications and Mobile Computing*, vol. 2019, pp. 1–9, 01 2019.
44. S. Jiong, L. Jin, J. Li, and Z. Fang, "*A smart parking system based on NB-IoT and third-party payment platform,*" *2017 17th International Symposium on Communications and Information Technologies (ISCIT)*, Cairns, QLD, Australia, pp. 1–5, 09 2017.
45. M. Kais, E. Bajic, F. Chaxel, and F. Meyer, "Overview of cellular LPWAN technologies for IoT deployment: Sigfox, LoRaWAN, and NB-IoT," March 2018.
46. D. M. Hernandez, G. Peralta, L. Manero, R. Gomez, J. Bilbao, and C. Zubia, "*Energy and coverage study of LPWAN schemes for industry 4.0,*" in *2017 IEEE International Workshop of Electronics, Control, Measurement, Signals and their Application to Mechatronics (ECMSM)*, pp. 1–6, May 2017.
47. T. Janssen, M. Aernouts, R. Berkvens, and M. Weyn, "*Outdoor fingerprinting localization using SIGFOX,*" *2018 International Conference on Indoor Positioning and Indoor Navigation (IPIN)*, Nantes, France, pp. 1–6, 09 2018.

15 Localization Using Bat Algorithm in Wireless Sensor Network

Ramandeep Kaur and Gurpreet Singh Saini

St. Soldier Institute of Engineering and Technology,
Jalandhar, Punjab, India

CONTENTS

15.1 INTRODUCTION

The availability of a low-cost sensor having sensing and processing cost has made it possible for the diffusion of Wireless Sensor Network (WSNs) at the great magnitude that it is applied at every application these days [1–3]. WSNs consist of such autonomous nodes that are normally deployed over the areas where human reach is not feasible; hence, the sensor nodes perform this task of continuous data collection and help the user in taking required action [4,5]. Other than the battery of the sensor nodes, the location of the sensor nodes is also a significant concern. These nodes are deployed with the help of the aircraft, so in case any node devours its energy completely, its location has to be known so as to incorporate further action [6]. One such source that informs about the location is Global Positioning System (GPS); however, the cost involved in the installation of GPS over the various nodes makes it infeasible for the WSNs to operate with it. Nevertheless, to track the location of the unknown nodes, and hence, to design the localization algorithm, some of the nodes equipped with GPS are deployed in the network. These nodes are termed as beacon nodes.

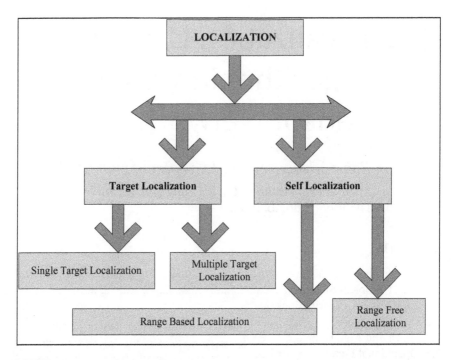

FIGURE 15.1 Localization techniques.

The process of determining the position of sensor nodes with the help of some beacon nodes is termed as localization [7]. Some of the nodes which are used for the detection of target nodes are termed as anchor nodes [8,9]. Since the development of WSNs, various localization algorithms have been designed [10–13]. The various techniques of localization are shown in Figure 15.1. While considering the literature survey carried out so far, it is observed that localization of the nodes is essential. Through the extensive research study, the following problems are identified.

In one of the methods of localization, the anchor node is made to move in the network to localize the target nodes [14]. The anchor node is moved in the Hilbert Curve trajectory, and this trajectory is moved through the different optimization techniques [15,16]; however, it is observed that the Bat algorithm is not yet explored for this purpose. It is further observed through the literature study that the use of virtual anchor nodes is exploited that helps in locating the target nodes. However, the recent work included six virtual anchor nodes, which increases the complexity of the network and also decreases the convergence for the optimization technique being employed.

15.1.1 Problem Statement

Once the nodes are deployed in the network, the concern that is raised is related to their location. Installing GPS on the sensor modules brings huge cost to the network, however, to avoid this expense of the network, the localization algorithms need to be developed. Although various optimization techniques have been introduced to handle the localization concern, there is still scope for the optimal localization of the sensor nodes. The existing algorithms either are very low in convergence or are not delivering the satisfactory performance. Therefore, it is observed that there are research efforts missing that deals with the localization algorithm who are having high convergence and also the least localization error.

15.1.2 MAJOR CONTRIBUTIONS

The major contributions of our work are stated as follow.

a. In this work, we propose Localization using the Bat algorithm (BA) i.e., LBA in which Hilbert-curve trajectory is used for locating the target nodes. This trajectory is used by the moving anchor nodes.
b. LBA employs three anchor nodes, of which only one is equipped with a GPS system.
c. Finally, the performance validation of the proposed work is done against the existing localization algorithms.

The rest of the chapter is organized as follows. Section 15.2 discusses the related work. Section 15.3 gives the detailed working of LBA. Results and simulation are discussed in Section 15.4. Finally, the conclusion is presented in Section 15.5.

15.2 LITERATURE WORK

The advancement in the field of WSN has focused on enhancing the energy efficiency of wireless sensor nodes, various such studies are available in the existing literature [17–23]. Since the development of WSN, the localization techniques have been helping in locating the randomly deployed sensor nodes. Liu et al. [24] proposed a virtual anchor node-based localization, which figures out the highly precise location of the unknown nodes and then upgrades them as a virtual node for the purpose of localization [25]. Euclidian and DV hop with VANLA are evaluated for the precision of the localization. The precision and the cost of localization are important criteria, as effectiveness is determined based on that information. Since the development of WSN, the localization in WSN has been the prominent topic of research [26–38]. Table 15.1 discusses various studies related to the localization techniques.

15.2.1 BAT ALGORITHM

The Bat optimization algorithm was developed by Xin-She Yang in 2010 [37]. Bats are fascinating creatures [40]. They are the only mammals having wings and an innovative skill of finding location according to sound, called echolocation, as shown in Figure 15.2.

Bats utilize echolocation to a definite angle; from all the types, micro-bats utilize more echolocation as compared to mega-bats. Micro-bats utilize echolocation to find food, evade hurdles, and discover its resting cracks in the night. The Bat algorithm was established depending upon the echolocation process of bats. In this process,

TABLE 15.1
Study of Various Localization Methods

Study Reference	Name of Method Employed	Types	Main Objectives	Homo/Hetero	Outcomes	Research Gap
He et al. (2003) [26]	APIT algorithm	Range free	— For multitude and dependent application	Homogenous	— Reduced effect of location error on routing	
Liu et al. (2004) [27]	Ring Overlapping based on comparison of Received Signal Strength Indicator	Range free	— Small intersection area and results in accurate location estimation	Homogenous	— Robust under irregular radio propagation patterns	
Tian et al. (2007) [28]	Selective anchor node localization algorithm (SANLA)	Range free	— More precision to execute localized process	Homogenous	— Good localization accuracy	Increasing the cost
Zanca et al. (2008) [29]	ML, Min-Max, Multilateration and ROCRSSI	Range based	— To acquire limits of localization algorithms	Homogenous	— Reliability, performance and localization errors	Line of sight problem
Kumar et al. (2012) [30]	LBA and BBO	Range based	— No of nodes localized — Accuracy computation time		— Faster — Matured and accurate localization	Multi-hop localization
Sabale et al. (2017) [31]	D-connect	Range based	— Novel path planning — Localization with minimum trajectory	Homogenous	— Better location estimation — localization error	Energy efficiency
Singh et al. (2017) [32]	PSO, LBA, BBO, FA optimization on umbrella projection	Range based	— Problem of LOS — Flip ambiguity	Heterogeneous	— Lower convergence rate — Good accuracy	Impact of flip ambiguity

(Continued)

TABLE 15.1 (Continued)
Study of Various Localization Methods

Study Reference	Name of Method Employed	Types	Main Objectives	Homo/Hetero	Outcomes	Research Gap
Singh et al. (2018) [33]	Optimization of VAN by PSO, LBA, BBO, FA	Range based	— Accuracy — Scalability	Homogenous	— Minimized LOS by VAN — Scalability	More accuracy Centralized
Tuba et al. (2018) [34]	Swarm intelligence, Firefly algorithm, trilateration	Range based	— Localization error.	Homogenous	— Improves accuracy — Low localization error	Line of sight problem
Phoemphon et al.(2018) [35]	Fine tune technique with Fuzzy based centroid localization algo	Range free	— Location accuracy — Coverage with help of VAN	Homogenous	— High estimation accuracy — Low Mean localization error	Efficient energy protocol
Shit et al. (2018) [36]	Directionality based algo, terrain and hop-terrain	Range based	— Localization error — Improve accuracy	Homogenous	— Scalability — Defined and dynamic path covering.	System with mobility and capability of interaction.
Kim et al. (2018) [37]	Threshold-Weighted centralized location algorithm with intersection threshold inconsideration of path loss	Range based	— Improving accuracy and complexity in small scale fading	Homogenous	— Cost effective with reduce no of anchor nodes	Mismatching of distance between unknown and test node.
Liu et al. (2018) [38]	Virtual Anchor-Based Localization Algorithm	Range	— Improve localization accuracy and cost effectiveness	Homogenous	— Low localization error	Enhancing precision of anchor nodes

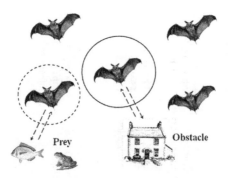

FIGURE 15.2   Real behavior of bats. [38]

pulses are created by bats that stay alive for 8–10 ms at some constant frequency. Some significant characteristics of the Bat algorithm are as follows.

a. Non-requirement of visibility to sense and estimate the distance for the food and the objects.
b. Association with velocity, position, and frequency with changing loudness and wavelengths.
c. The value of loudness is dependent upon the kind of strategies employed.

BA initiates with some random initial population of bats in which the position of the bat i represented by x^t and velocity is denoted by v^t at time t. x_i^{t+1} and v_i^{t+1} denote the new position and new velocity of bat i at time $t + 1$. The α_{min} and α_{max} denote the minimum and maximum values of pulse frequency, respectively. β is random number that varies from [0 1].

$$\alpha_i = \alpha_{min} + \left(\alpha_{max} - \alpha_{min}\right) \times \beta \tag{15.1}$$

$$v_i^{t+1} = v_i^t + \left(x_i^t - x^{best}\right) \times \alpha \tag{15.2}$$

$$x_i^{t+1} = x_i^t + v_i^{t+1} \tag{15.3}$$

15.3   OPERATIONAL FUNCTIONING OF LBA

In this chapter, the simulation analysis is done in MATLAB® software. The performance metrics are used here to evaluate the performance of proposed protocol against the HPSO- and PSO-based protocol.

15.3.1   NETWORK MODEL

The realization of the proposed work is followed by the network assumptions which should be addressed. These are mentioned as follows.

a. The nodes are deployed randomly in the network in the given network area [39].

b. The target nodes are those nodes whose location is to be determined.

c. The anchor node is a mobile node whose initial position is decided by the user.

d. In this work, the mobile anchor node is made to move in a circular region rather than Hilbert curve move as done in the existing algorithm.

e. The localization of anchor nodes is possible only if they happen to be in the vicinity of the mobile anchor node.

f. The use of virtual nodes is only for performing localization of the sensor nodes.

g. The movement of mobile anchor is done in a circular fashion as it updates the position along the circle.

h. The presence of physical objects is not considered.

i. Nodes are location unaware despite some of the nodes which are equipped with the GPS modules.

j. The placement of virtual nodes is done with the angle of 60 degrees.

15.3.2 Simulation Parameters

The simulation analysis is done based on the parameters covered in Table 15.2. These parameters are considered for simulation in MATLAB® Software.

15.4 RESULTS AND DISCUSSION: PERFORMANCE EVALUATING METRICS

We have considered two metrics to evaluate the performance of the proposed work in comparison to the existing PSO-based and HPSO-based protocol targeting localization.

a. *Convergence time*: When the optimization technique is processed to deliver an optimized solution, the time it has taken to do so is termed as convergence time. While performing for various harsh applications, it proves to be a crucial

TABLE 15.2
Simulation Parameters for LBA and Network Scenario

Simulation Parameters	Value
Number of target nodes	50
Size of population	20
Total iteration	100
Inertia weight	0.729
Value of cognitive learning parameter	1.494
Social learning parameter	1.494
Random values r_1 and r_2	[0 1]
Noise variance	0.1
Network area	15×15 square meter
Total virtual nodes	2
Angle	60 degree

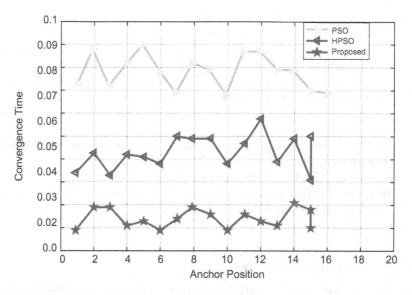

FIGURE 15.3   Convergence time vs anchor position.

parameter. As shown in Figure 15.3, the proposed technique has more convergence time in comparison to the other techniques.
  b. *Localization error*: To acquire the main objective of reducing the localization
     error for the proposed technique, the proposed technique has performed exclu-
     sively well as compared to the competitive protocols as shown in Figure 15.4.
     It is observed that the proposed method has a localization error of less than 0.2,
     whereas the value for the same is higher for the other protocols.

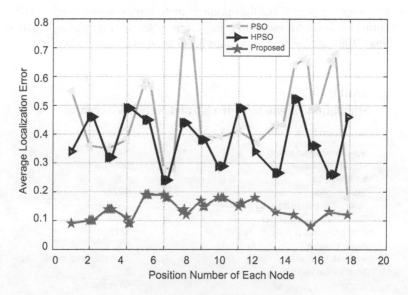

FIGURE 15.4   Localization error vs. position number of each node.

15.5 SUMMARY

It can be said that proposed technique has outperformed the HPSO- and PSO-based localization techniques in terms of the two-performance metrics, namely, convergence time and localization error. Movement of anchor nodes in a circular way not only covers the sensing area of the whole network, but also helps to cover the randomly deployed nodes in the network. In the previous cases of HPSO, the main focus had been on the Hilbert curve movement which is restricted in the proposed work as it had been covering those areas as well where the sensor nodes might not have been deployed.

15.6 CONCLUSION

The localization of the sensor nodes once they are deployed in the network area is of utmost concern. It ensures the precise location estimation of the event-triggered area. While doing so, the rescue operation can be taken in lieu of any catastrophe events. In this paper, we have used the Bat algorithm to reduce the localization error in the network. One anchor node is made to move in the Hilbert-curve path to localize the target nodes using the Bat algorithm. The distance estimation, position estimation and finally the localization error is computed. It is observed that the proposed LBA outperforms various other algorithms comprehensively in the context of convergence time and localization error. This technique is highly suitable for those critical applications where the location of the event has to be determined precisely.

REFERENCES

1. I. F. Akyildiz, W. Su, Y. Sankarasubramaniam, and E. Cayirci, "A survey on sensor networks," *IEEE Communications Magazine*, vol. 40, no. 8, pp. 102–114, 2002.
2. B. M. Sahoo, H. M. Pandey, and T. Amgoth. "GAPSO-H: A hybrid approach towards optimizing the cluster based routing in wireless sensor network." *Swarm and Evolutionary Computation*, vol. 60, p. 100772, 2020.
3. S. Verma, N. Sood, and A. K. Sharma, "A novelistic approach for energy efficient routing using single and multiple data sinks in heterogeneous wireless sensor network," *Peer-to-Peer Networking and Applications*, vol. 12, no. 5, pp. 1110–1136, 2019.
4. B. M. Sahoo, R. K. Rout, S. Umer, and H. M. Pandey. "*ANT Colony Optimization based optimal path Selection and data gathering in WSN.*" In *2020 International Conference on Computation, Automation and Knowledge Management (ICCAKM)*, pp. 113–119. IEEE, 2020.
5. B. M. Sahoo, T. Amgoth, and H. M. Pandey. "Particle swarm optimization based energy efficient clustering and sink mobility in heterogeneous wireless sensor network." *Ad Hoc Networks*, vol. 106, p. 102237, 2020.
6. J. Wang, R. K. Ghosh, and S. K. Das, "A survey on sensor localization," *Journal of Control Theory and Applications*, vol. 8, no. 1, pp. 2–11, 2010.
7. M. Rudafshani and S. Datta, "*Localization in wireless sensor networks*," In *Information Processing in Sensor Networks, 2007. IPSN 2007. 6th International Symposium on*, pp. 51–60, 2007. Online. Available: http://ieeexplore.ieee.org/abstract/document/4379664/ (accessed September 18, 2017).
8. H. Fang, X. Cui, and Q. Liu, "Review of the localization problem in wireless sensor networks J," *Computer and Information Technology*, vol. 6, 2005.

9. P. Y. W. Dan, "A review: Wireless sensor networks localization," *Journal of Electronic Measurement and Instrument*, vol. 5, 2011.

10. A. Boukerche, H. A. Oliveira, E. F. Nakamura, and A. A. Loureiro, "Localization systems for wireless sensor networks," *IEEE Wireless Communications*, vol. 14, no. 6, pp. 6–12, 2007.

11. P. Rong and M. L. Sichitiu, "*Angle of arrival localization for wireless sensor networks,*" In *2006 3rd Annual IEEE Communications Society on Sensor and Ad Hoc Communications and Networks*, 2006, vol. 1, pp. 374–382.

12. G. Mao, B. Fidan, and B. D. Anderson, "Wireless sensor network localization techniques," *Computer Networks*, vol. 51, no. 10, pp. 2529–2553, 2007.

13. D. A. Tran and T. Nguyen, "Localization in wireless sensor networks based on support vector machines," *IEEE Transactions on Parallel and Distributed Systems*, vol. 19, no. 7, pp. 981–994, 2008.

14. P. Singh, A. Khosla, A. Kumar, and M. Khosla, "Optimized localization of target nodes using single mobile anchor node in wireless sensor network," *AEU—International Journal of Electronics and Communications*, vol. 91, pp. 55–65, 2018.

15. S. Kaur, L. K. Awasthi, and A. L. Sangal, "HMOSHSSA: A hybrid meta-heuristic approach for solving constrained optimization problems," *Engineering with Computers*, pp. 1–37, 2020.

16. S. Kaur, L. K. Awasthi, A. L. Sangal, and G. Dhiman, "Tunicate Swarm Algorithm: A new bio-inspired based metaheuristic paradigm for global optimization," *Engineering Applications of Artificial Intelligence*, vol. 90, p. 103541, 2020.

17. S. Verma, R. Mehta, D. Sharma, and K. Sharma, "Wireless sensor network and hierarchical routing protocols: A review," *International Journal of Computer Trends and Technology (IJCTT)*, vol. 4, no. 8, pp. 2411–2416, 2013.

18. S. Verma, S. Kaur, A. K. Sharma, A. Kathuria, and M. J. Piran, "Dual sink-based optimized sensing for intelligent transportation systems," *IEEE Sensors Journal*, 2020 doi:10.1109/JSEN.2020.3012478.

19. S. Verma, N. Sood, and A. K. Sharma, "QoS provisioning-based routing protocols using multiple data sink in IoT-based WSN," *Modern Physics Letters A*, vol. 34, no. 29, p. 1950235, 2019.

20. S. R. Pokhrel, S. Verma, S. Garg, A. K. Sharma, and J. Choi, "An efficient clustering framework for massive sensor networking in industrial IoT," *IEEE Transactions on Industrial Informatics*, vol. 17, no. 7, pp. 4917–4924, 2020.

21. S. Verma, N. Sood, and A. K. Sharma, "Cost-effective cluster-based energy efficient routing for green wireless sensor network," *Recent Advances in Computer Science and Communications*, December 31, 1969. http://www.eurekaselect.com/180151/article (accessed April 17, 2020).

22. S. Verma and K. Sharma, "Zone divisional network with double cluster head for effective communication in WSN," *International Journal of Computer Trends and Technology*, vol. 4, no. 5, pp. 1020–1022, 2013.

23. D. Pant, S. Verma, and P. Dhuliya, "*A study on disaster detection and management using WSN in Himalayan region of Uttarakhand,*" in *2017 3rd International Conference on Advances in Computing, Communication & Automation (ICACCA)(Fall)*, 2017, pp. 1–6.

24. C. Liu, K. Wu, and T. He, "*Sensor localization with ring overlapping based on comparison of received signal strength indicator,*" in *2004 IEEE International Conference on Mobile Ad-hoc and Sensor Systems (IEEE Cat. No. 04EX975)*, 2004, pp. 516–518, Fort Lauderdale, FL, USA.

25. G. Sharma, A. Kumar, P. Singh, and M. J. Hafeez, *"Localization in wireless sensor networks using invasive weed optimization based on fuzzy logic system,"* in *Advanced Computing and Communication Technologies*, Springer, 2018, pp. 245–255.
26. T. He, C. Huang, B. M. Blum, J. A. Stankovic, and T. Abdelzaher, *"Range-free localization schemes for large scale sensor networks,"* in *Proceedings of the 9th Annual International Conference on Mobile Computing and Networking*, 2003, pp. 81–95, San Diego, CA, USA.
27. S. Tian, X. Zhang, X. Wang, P. Sun, and H. Zhang, *"A selective anchor node localization algorithm for wireless sensor networks,"* in *2007 International Conference on Convergence Information Technology (ICCIT 2007)*, 2007, pp. 358–362, Gwangju, Korea (South).
28. G. Zanca, F. Zorzi, A. Zanella, and M. Zorzi, *"Experimental comparison of RSSI-based localization algorithms for indoor wireless sensor networks,"* in *Proceedings of the Workshop on Real-World Wireless Sensor Networks*, 2008, pp. 1–5, New York, NY, USA.
29. A. Kumar, A. Khosla, J. S. Saini, and S. Singh, *"Stochastic algorithms for 3D node localization in anisotropic wireless sensor networks,"* in *Proceedings of Seventh International Conference on Bio-Inspired Computing: Theories and Applications (BIC-TA 2012)*, 2013, vol. 1, pp. 1–14.
30. K. Sabale and S. Mini, "Anchor node path planning for localization in wireless sensor networks," *Wireless Networks*, vol. 25, no. 1, pp. 49–61, 2019.
31. P. Singh, A. Khosla, A. Kumar, and M. Khosla, "3D localization of moving target nodes using single anchor node in anisotropic wireless sensor networks," *AEU – International Journal of Electronics and Communications*, vol. 82, pp. 543–552, 2017.
32. E. Tuba, M. Tuba, and M. Beko, "Two stage wireless sensor node localization using firefly algorithm," in X.-S. Yang, A. K. Nagar, and A. Joshi (Eds.), *Smart Trends in Systems, Security and Sustainability*, Springer, 2018, vol. 18, pp. 113–120.
33. S. Phoemphon, C. So-In, and D. T. Niyato, "A hybrid model using fuzzy logic and an extreme learning machine with vector particle swarm optimization for wireless sensor network localization," *Applied Soft Computing*, vol. 65, pp. 101–120, 2018.
34. R. C. Shit, S. Sharma, D. Puthal, and A. Y. Zomaya, "Location of Things (LoT): A review and taxonomy of sensors localization in IoT infrastructure," *IEEE Communications Surveys and Tutorials*, vol. 20, no. 3, pp. 2028–2061, 2018.
35. K.-Y. Kim and Y. Shin, "A distance boundary with virtual nodes for the weighted centroid localization algorithm," *Sensors*, vol. 18, no. 4, p. 1054, 2018.
36. P. Liu, X. Zhang, S. Tian, Z. Zhao, and P. Sun, *"A novel virtual anchor node-based localization algorithm for wireless sensor networks,"* in *Sixth International Conference on Networking (ICN'07)*, 2007, pp. 9–9, Sainte Luce, Martinique, France.
37. X.-S. Yang, *"A new metaheuristic bat-inspired algorithm,"* in *Nature Inspired Cooperative Strategies for Optimization (NICSO 2010)*, Springer, 2010, pp. 65–74.
38. S. Amri, F. Khelifi, A. Bradai, A. Rachedi, M. L. Kaddachi, and M. Atri, "A new fuzzy logic based node localization mechanism for wireless sensor networks," *Future Generation Computer Systems*, vol. 93, pp. 799–813, 2017.
39. S. Verma, S. Kaur, M. A. Khan, and P. S. Sehdev, "Towards green communication in 6G-enabled massive internet of things," *IEEE Internet of Things Journal*, vol. 8, no. 7, pp. 5408–5415, 2020, doi:10.1109/JIOT.2020.3038804.

Index